PRAISE FOR *While Glaci*

"Through Jackson's lyrical prose, readers
of both human relationships and glacial landscapes, which,
Jackson reminds us in one of the book's most eloquent
passages, are made up of 'millions and millions of little
snowflakes, reaching out to one another, grasping hands.'"

—Stephen Siperstein, *The Goose*

"Climate change, she convinces us, is not just about science—
it is also about the audacity of human courage
and imagination."

—*Yale Climate Connections*

"By humanizing the glaciers, Jackson brings this natural
phenomenon to a more personal level."

—*Seattle Spectator*

"Climate change is many things, including an upheaval—
sudden and violent—in the life of our planet. As such, it
unleashes feelings and forces like those in a family when
someone dies. This is a profound way of thinking about
where we are right now, and what we better do about it."

—**Bill McKibben**, founder of 350.org
and author of *Eaarth* and *The End of Nature*

"As the poet Tony Hoagland has pointed out, most of us '…
walk like zombies through our burning dying world…'
Not so M Jackson, who moves through the world very
much aware of both the little and individually important
things, such as family, while simultaneously perceiving and
understanding the catastrophe that is happening all around
us. In *While Glaciers Slept,* she links the one to the other in a
flawless and brilliant way. This is superb."

—**Carlos Martinez**, author of
The Cold Music of the Ocean and
The Raw Silk of the Dark

"Jackson, a National Geographic Expert and prominent scientist passionate about researching glacial systems, explores in this emotional memoir her experience of losing her parents, one after the other, to cancer. Literally and metaphorically, the author compares the hopelessness she felt in the aftermath of their deaths with the depression people sometimes encounter witnessing the destruction of the environment. While at times on the verge of giving up in the face of such personal upheaval, Jackson persevered in learning a new way of living, as humanity will have to do with the advent of climate change. She offers parallel glimpses of optimism, both for herself and for the future of the planet, sharing her journey of growth and discovery while at the same time highlighting imaginative, radical projects proposed by innovative thinkers designed to avert what most scientists believe to be inevitable: a changed earth.

Reminiscent of Bill McKibben's *Eaarth*, this title will interest readers of environmental issues, particularly climate change and a warming Arctic region, and fans of personal narratives."

—*Library Journal (Starred Review)*

PRAISE FOR *The Secret Lives of Glaciers*

"When it comes to glaciers, Dr. M Jackson is a linguistic sorcerer, making you fall in love by proxy with the geological memory-keepers. . . . Jackson's text moves with historical and scientific precision to show that human beings and glaciers exist in reciprocal relationship to one another, and have done so across epochs of glacial progressions and retractions. . . . Glaciers speak to our future just as much as we speak to theirs, and M Jackson's epic examination of their place in humanity's story is compelling."

—*Foreword Reviews (Starred Review)*

"The most compelling narratives are Jackson's own"
—*Anchorage Daily News*

"This outrageous book, rich with revelation and stewardship, is, at its deepest level, an icy blue love story to make us reconsider what it means to be fully alive—and open to wonder—in our ever-changing world."

—Kim Heacox, author of *John Muir and the Ice That Started a Fire*

"M Jackson brings a powerful combination of skills to bear in her ambitious task of complicating our understanding of the rapidly dwindling masses of ice with which we share this planet. Blending hands-on science, vivid descriptive writing, affecting personal anecdote, and insightful cultural observation, *The Secret Lives of Glaciers* is a hypnotic and inspiring book—essential reading for anyone who loves nature and is concerned about the human species' continued existence within it."

—Tim Weed, author of *A Field Guide to Murder & Fly Fishing*

"M Jackson is a master storyteller, weaving evocative anecdotes and historical and scientific narratives into an intricate dance of the relationship between man and ice. Jackson writes eloquently, her stories of the real, concrete effects of climate change on the people of Iceland both informative and heart-wrenching."

—Dr. Michele Koppes, Glaciologist and Geographer

"*The Secret Lives of Glaciers* engulfs you from the very first page, and in that way does due justice to the colossal yet fragile icy protagonist it intends to uncover for its readers."

—Asher Jay, Conservationist and National Geographic Explorer

"It seems safe to say that much of the literature around glaciers and climate change can be a little dry—no pun intended. Scientific texts on mass balance, false ogives, ground lines, dendrochronology and the cryosphere can be a little heavy for the glacier-curious layman. This isn't the case with *The Secret Lives of Glaciers*, a newly published book by American geographer, glaciologist and National Geographic Explorer M Jackson. The book takes the unusual tack of reporting climate change as a series of stories told by M and the people she meets during her time spent researching glaciers in Höfn. Containing elements of autobiography and diaristic accounts of the glaciers alongside conversations, observations, and anecdotes of all kinds, it's approachable and readable stuff."

—John Rogers, *The Reykjavík Grapevine*

"In [*The Secret Lives of Glaciers*] we learn how glacial retreat is impacting communities, the connection between extractive tourism, extractive science, and glaciers, why it matters that the majority of glaciology has been produced by white men, and the ways in which polar and mountain explorations have furthered colonial, capitalist, and imperialist projects.... *The Secret Lives of Glaciers* explores the heartfelt connections between people, place, and ice."

—*For the Wild (interview with Ayana Young)*

while glaciers slept

WHILE
GLACIERS

GREEN WRITERS PRESS *Brattleboro, Vermont*

SLEPT

Being Human in a Time of Climate Change

M Jackson

Foreword by Bill McKibben

Originally published in hardcover by Green Writers Press in 2015.

Author's note: As this book spans many years and several continents, I have had to re-imagine events and conversations based on occasionally incomplete memories in my mind. I have tried to faithfully represent the spirit of those scenes within these pages. For legal and ethical reasons, certain names and episodes have been changed.

Printed in the United States of America

10 9 8 7 6 5 4 3 2 1

Green Writers Press is a Vermont-based publisher whose mission is to spread a message of hope and renewal through the words and images we publish. Throughout, we will adhere to our commitment to preserving and protecting the natural resources of the earth. To that end, a percentage of our proceeds will be donated to environmental activist organizations.

GReen
wriTers
press

Giving voice to writers and artists who will make the world a better place
Green Writers Press | Brattleboro, Vermont
www.greenwriterspress.com

LIBRARY OF CONGRESS CONTROL NUMBER: 2015936482
ISBN: 978-0-9968973-3-4

Photographs courtesy of: Colin Aiken, Kyle Dungan, David Estrada, Grant Jackson, M Jackson, Sarah Jackson, the Jackson Family Collection, Jon Marshall, Federico Pardo, and Elizabeth Ruff.

Author photograph by Annie Agnone.
Cover design by Ani Pendergast.
Book design by Dede Cummings and Ani Pendergast. Set in Bembo and Frutiger.

Visit the author's website for information about her work in Iceland, booking her for a talk, discussion guides, etc., at www.drmjackson.com.

This book is dedicated to Sarah and Grant,
who keep my compass pointed home.

Any landscape is a condition of the spirit.

—HENRI FREDERIC AMIEL

"But I don't want to go among mad people," Alice
  remarked.
"Oh, you can't help that," said the Cat. "We're all mad
  here. I'm mad. You're mad."
"How do you know I'm mad?" said Alice.
"You must be," said the Cat, "or you wouldn't have
  come here."

—LEWIS CARROLL

while glaciers slept

Foreword | Bill McKibben

"I think our planet is slowly becoming disabled. Due to climate change, in natural processes that digest carbon, regulate temperature, keep the climate on an even keel, everything is off kilter. Undoubtedly, unquestionably, it has been shown with certainty: the way we live our lives is causing these systems to fail." —M JACKSON

THIS IS TRUE—AS TRUE AS IT IS POSSIBLE TO GET. There is a huge elephant always present in every room on our planet right now, the elephant of climate change. Nothing humans have ever done is so big, and nothing so big has ever been so thoroughly ignored.

And the reasons it's been ignored, and the reasons we must ignore it no longer, are the reasons in this book. Not, ultimately, the prosaic and practical questions about sea level rise and increased risk of drought and ocean acidification. These things are all crucially important, but they're not the core. At the core, somehow, is the question of whether the big brain was a good adaptation.

Or, more precisely, if it came attached to a big enough heart to get us out of the trouble we're in. To get us out of the habit of staring at the shiny object nearest by, and to look instead at the mountain, the forest, my wife, your mother, our meaning. Those are the kinds of questions easiest to answer in the company of caribou and humpback, or of family and friends. The real company, not the virtual, pretend, screen-based company. We live in an abstracted, mediated world, and in that kind of world it seems possible that all that is real and beautiful might slip right by us—especially our home planet in all its buzzing, complex, cruel glory.

And so we fight. Sure we screw in the new lightbulb, but mostly we screw up our courage. Screw up our courage to well and truly love. That's what this book is about, I think; I hope you read it in the spirit of openness it deserves, making yourself vulnerable to both hurt and joy. We may or may not be able to slow down climate change (I hope we are able, and so I devote my days to that task). But we are definitely able to witness the world, and ourselves on it, in these fragile and lovely moments. That's our task, too.

—Bill McKibben

O<small>N MY DESK TODAY SITS A LETTER I FOUND WRITTEN</small> by my dad to my mom. A year ago, my brother, my sister, and I were sorting through the old desk at the farm. It was the first step, the first of packing up, cleaning out, choosing what was important and what wasn't. We chose to start with the desk because it wasn't their bedroom, it wasn't the loft, it wasn't the kitchen, it wasn't all those other places where the fingerprints of our parents rained thick and deep. We thought, then, that the desk was full of business papers and insurance forms and bank statements. Neutral things. Not the things that really make up a person. But—

It also contained a letter. Buried under stacks of manila envelopes and plastic cases and paperclips and thumbtacks.

Undated, on yellowed paper, with no envelope: my dad's distinctive engineering script. I saw the handwriting before I saw the letter, before I registered it for what it was. Each capital *I* looked like a *J*. There was a faint stamp on the upper left corner—perhaps this was a postcard sent through the mail. It read:

Dear Madam.

I declare my love for you, and wish to beg you to consider leaving your husband and running away

with me to an enchanted deserted island. If you say
no, I shall hurl myself over the Narrow's Bridge.

I love you.
John

I've framed this letter, and it sits on my desk.

My parents are dead. They both had terminal cancers,
and they passed away within two years of each other. I was
twenty-six.

Those are the basics, but there is much more.

All stories have so much more.

My mom was married to someone else when she met
my dad. But at the time the letter was written my parents
were married. I know this because my dad is humorously
threatening to jump off the Narrows Bridge in Puget Sound.
My parents moved to Washington State several years into
their relationship, years after Mom had divorced her previous
husband and began to build a life with my dad. They'd lived
all over the map, traveling north and south in bell-bottomed
jeans along twisty roads in unreliable vehicles, searching always for a place to believe could be home.

And then, lifetimes before they found their real home,
they settled for a couple of years underneath the Narrows
Bridge in Washington State on a small scrap of land called
Salmon Beach. From the deck of their tiny house on stilts
they could look out across the water and see the boat traffic
under the bridge and the car traffic over it. The bridge was
part of their daily viewscape, part of their daily conversation,
just like today, at the farm that eventually became their home,
the view of Mt. Rainier or the surrounding fields is part of
the daily fabric of our lives.

The letter was probably a note of whimsy, a single moment between two people that holds details long lost now.
But at its core it was a reminder from my dad to my mother

that he loved her. He loved her right through her death, and certainly through his.

Everything I knew and trusted in this world changed when my parents became terminally ill. The foundation of the surrounding landscape, the regularity of the weather, the creep of the sun's light over the spinning planet—all these and more shifted simultaneously by incremental and astronomical degrees. And even though I told myself the worst would not happen, even though I denied it until the very last moment, even though I made plans for after and said good-bye, I did not believe until it happened.

This story, however, is not just about death—their deaths, that moment when cancer outbalanced life in their bodies—nor is it about sadness, or irretrievable loss. Rather, this is a story that contains such multitudes. Here is the story of two people facing unknown futures and their daughter's trying to recognize again a landscape and lifescape transformed. And here also is the story of all of us and our single and shared experiences as human people across this planet in the midst of the greatest of changes.

In 2007, when I was twenty-five, I lived in Africa, up in the far northwestern corner of Zambia, serving as a Peace Corps Volunteer. This was two years before my mother died. I resided in a small, mud-bricked, grass-thatched hut perched on a ridge above a wide, shallow valley. Threading along the floor of the valley was a narrow stream bordered by thick grasses and sedges. Most evenings, I walked the forty-minute footpath down the valley floor and along the stream until I reached an opening in the grasses along the bank. There, I'd slip through the vegetation and slide down the red clay to the water's edge. A large rock nosed up through the water and reeds. I always placed my soap, shampoo, and clothes across its gray face: my built-in river table.

I discovered the stream several months after moving into my hut. My discovery was jubilant. Previously, water collection involved digging a hole in the mud-swamp, waiting for the water to seep into it, dunking a bucket into the hole, and then carrying the buckets up the hill to my hut. A round trip for one bucket took half an hour. Every time I had wanted to take a bath, I needed three buckets of water.

Water was the first way I connected with the Kaonde villagers I lived with. The Kaonde are a tribe of Zambian people living across the North-Western Province of the country. Those Kaonde who lived in the rural area where I was assigned as a Peace Corps Volunteer practiced rain-fed agriculture: they were dependent on the mercurial skies to nourish their crops of maize, pumpkins, and greens. If it did not rain, many Kaonde farmers in my area—men, women, and children—hand-watered crops with heavy buckets that were carried on heads and shoulders and hips and backs and walked up and down and up and down the many, many rows in the fields.

Increasingly, the farmers spoke of changes in the rain. They told stories of various years, when the rainy season arrived at unusual times, and heavier, or lighter, or just different. So much rain, so quickly, so soon, destroyed the tender shoots of maize and flooded the fields, resulting in even more ample breeding grounds for malarial mosquitoes. And then, unpredictably, the rainy season would end, like someone jammed a cork straight into the cloudy heavens, right when the maize or pumpkins or greens needed that last burst of water, it was withheld, and the buckets came out. The Kaonde didn't have the language for climate change.

During the dry season, I spent hours each day hauling water, balancing old, yellow buckets against my chest and waddling up the path to my hut, visions of laundry and dishes and baths crowding each step I took. Village women, my friends at any other point in the day, would shriek and laugh because I did not have the neck strength to balance forty-five pounds of water on my head.

Finding a small bathing area in the stream where I could submerge my body changed the quality of my life in Zambia.

I spent long evenings just floating, weightless, feeling the soft tendrils of currents whispering over my tired muscles. My hair and the watery weeds would mingle, twining, and my body would dissolve layer by layer into the murky water. After a few months, it felt like the only place I could really think was in that stream with my head half-submerged. Watching the clouds billow overhead, I could fill my eyes with the sensory overload of primary colors. The ground was red, the vegetation green, the sky blue, the clouds white. Bright yellow weaver birds danced and chirped, threading nests together on tiny reeds bent over the water.

Once, I floated past my bathing station and downstream. I drifted along on my back, twisting my fingers in the reeds as they floated past. I have no sense of how long I meandered, gently bobbing, before my head lightly bumped into a downed acacia. Even though the tree had almost snapped over, it still carried on with its tree business, alive and well in the Zambian evening.

I remember rising out of the water and looking around. I was surprised to sight a small pond the stream fed. I'd floated a lot farther downstream than I'd ever hiked before. Finally, I was in a place where I could actually swim.

The stream in which I bathed daily was about five feet across, with large, soft, marshy shoulders crowding into the water and reducing the actual reach of the stream to about three feet. Here, this new pond was long and narrow, about fifteen feet across, but felt to me wide, opulent. The water was clear—surprisingly so. I stroked out into the middle of the pond and dived down. It was deep. Such luxury made me love the rainy season, the time of plentiful water.

I hit bottom and guessed it to be about fifteen feet down. It was muddy and grassy. Rotating my body so I was sitting on the bottom of the lake, I remember looking up at the surface. My long hair waved in and out of my sight. The blues and whites and greens blended. I opened my mouth and large, perfectly clear and round bubbles escaped. I followed their journey as they twirled and danced, floating skyward. As soon as they hit the surface, they broke and disappeared.

I remember when the Intergovernmental Panel on Climate Change (IPCC) released its fourth report in 2007. It took six years to produce, and its tone was solemn, almost in awe of its own verdict. In language as clear as day, it stated to anyone who would read it: "Warming of the climate system is unequivocal."

"Act now," urged the thousands of people from across the planet who authored the report.

The "we" of the report was all-encompassing. It was not a message to the developing world, or to Europe, or to America. Rather, it pointedly said that this is humanity's challenge. Either reduce global greenhouse gas emissions, most notably carbon dioxide, or humanity begins to become undone on this planet. Climatic changes could arrive in the form of island nations submerged, a forty- to sixty-foot sea level rise, reductions in crop yields by 50 percent, mass species extinction, widespread glacier and ice cap recession.

"The time for doubt has passed," stated Dr. Rajendra Pachauri, chairman of the IPCC, after the report was released. "Slowing or even reversing the existing trends of global warming is the defining challenge of our ages. If there is no action before 2012, that's too late."

Climatic changes surround us completely. They seep into all aspects of the human and more-than-human world, invading society, knowledge, existence, trust. Every person today is experiencing these changes, and the culminating range of those experiences is shaped and whispered and narrated and decorated with phrases like "climate change," "global warming," "climate disruption," and "it doesn't snow like it used to."

It all seems so difficult to imagine, to surmount. But climate change's heart is small. It beats to the pulse of escalating greenhouse gas emissions, to widespread inaction and fear, to your and my being unable to imagine a future differently.

People say they can always find me by the sound of my bracelets clanking.

I wear silver bracelets in multitudes. Some live permanently on my right arm; others come out for particular reasons or occasions, in moments of triumph or success, in times of remembrance and sadness.

My bracelets are my personal conservation effort, my re-wilding of the landscape around my wrists. They are glacial remnants of the last ice age, they are refugia, and they are now the most immediate connection I have with my mother.

I have one bracelet that is always with me, though it is not worn or displayed often. It has three large, turquoise stones lined up oval-to-oval, with eight smaller stones orbiting. On its surface are constellations of nicks and dents, evidence of late evenings and sunshine, car doors slamming, arms crossed and tucked, movements and acts. It has been worn down, like a body.

When I hold this bracelet, heavy in my hand, it is suffused with warmth. There are whispers that accompany it, spectral inhabitants outside my peripheral vision, rattling, jangling. This bracelet has history—and company. Three smaller, silver bracelets clatter around it, a chatty family wearing away over time. When I wear them together, stacked on my wrist, my arm feels lighter, more supported.

I polish them in the morning and brood over each separately, clutching and holding, examining: I perch on the edge of a rickety bathtub and inspect how the silver survived the night.

There are others, too, additional bracelets, some wide, some not, one a rope of twisted silver, another crested with green turquoise. There is only one bracelet of the bunch that is truly mine: I bought it in Alaska years ago, when I first started working with glaciers, before I wore any bracelets. It reminded me then of my mother, of her bracelets singing on her arm as she tended the vegetable gardens back on our small family farm—far, far away from the ice field I worked on in Alaska.

All the others that I wear are from my mother. Over the years she gave me specific bracelets to mark specific events in my life. It was after she died, however, that I inherited the entire row of jangles she wore continuously, day after day, until her body wore down to dust. After her death, I removed them from the plastic hospital bag, held them, polished them, mourned them, and stacked them in a row on my wrist. They cinched down, clanking and clattering and acclimating to their new landscape, their new home. Their ends bit softly into my flesh, marking vague bruises reminiscent of my mother's purpura.

In her last years, my mother started wearing the huge turquoise bracelet every day. It became a permanent fixture on her right wrist. She wore it to access the strength and hope imbued in its soft center. She was given the bracelet as a young woman by my great-great-aunt Dot.

I have never felt the full pull of my ancestry, the need to know my blood. Until recently, I have been content with knowing my immediate family, which is small and easily countable on one hand. But of late, as I wade through the confusions of loss and death and heartbreak and that hand's countable fingers get twisted into a rather small fist, I find my daily strength replenished by the silver clanking on my arm. I am wearing the mycelial traces of the powerful women in my family, their lives and hopes and strengths and experiences overseen by clanking half-moons of witness, small, silver bracelets passed down generation by generation.

I remember, once, in Zambia, before I found my pond, I set off walking on a dusty afternoon. I went because I wanted to feel like I was actually doing something. I was listless and tired and focused on the giant sore mysteriously growing on my thigh, and I thought about the Rick Bass book I was reading and then my mind wandered and I thought of nothing else. So much of a typical Peace Corps experience is down time. I closed the flap to my hut, threw a log on the fire to keep it going, and strolled into the trees. I carried nothing.

The forest had the first few sips of water under its belt with the beginning of the rainy season. The early, teasingly small amount of water emboldened the primary colors, lending them a confidence, a hint of arrogance as they gouged and merged into one another.

At the time, the dirt had sucked up just enough moisture to make the ground look like a red that had given up on itself—dark, rusty pigment seeped everywhere underfoot, dyeing anything it touched, including my feet. My Chaco sandals and toes, my heels and ankles—a massacre had occurred down there, a muddy mess that worsened as I strolled along.

Walking was empowering to me, a physical act that bore fruit in the sense of accomplished motion. I never worried about getting lost—there was no actual place to be or time to tend—so in a sense, I existed in a perpetual state of arriving on time to every place I went. I first came to Zambia with clean clothes and high hopes, and I tried to navigate the arenas of time and place, tried to catch on to the local rhythm. Now, more experienced, I gave up on minutes, on seconds, on specific times.

One of the few concerns of wandering freely in the trees was the danger of rain. At my family's farm in Washington, rain was constant, lightly spitting from overcast, Northwest skies. Zambian raindrops were sudden, momentous, bone-breaking. Flooding was common.

About two, maybe three, hours into my walk, my forest eye-browsing was interrupted by soul-shaking claps of thunder. I recognized an immediate need for shelter—real shelter, not the huddle-at-a-tree-base shelter. Lightning and wind exploded and shrapnelled trees all the time, creating debris avalanches of branches and spears and all the crawly things that lived in the canopies. Crawly things terrified me.

I started to run, taking the twists and turns of the narrow bush path in leaps and bounds. This was not about getting wet.

The first few drops had just begun to crash down around me when I registered a grass thatched roof poking through the forest-green blur. A tiny path coursed off of the track I was on, leading towards the roof. I shot left.

It was a church, a small one, empty. Such structures were scattered throughout the area, handmade mud bricks stacked waist-high on three sides outlining spaces about ten feet wide by fifteen feet long. The bricks were stuck together with mud mortar and had a red sludge cemented smoothly over the floor inside. Four roughly-hewn poles held a roof, the bamboo latticing covered in thatch.

I ducked in as the skies boomed. As my eyes adjusted to the dim light, I recognized the basic jungle church setup: stools and benches in the form of rocks, small tree-stump bits, a few broken mud bricks scattered about the lower two-thirds of the floor. Up front, mud bricks were neatly elevated and arranged like a cross.

I claimed a tree stump and crouched down, the air chilling. Thunder boomed, depth charges detonating off the leeward side of the church. I could hear trees shaking, splitting. In less than two minutes, the rain sluiced off the thatch roof in a steady stream, wrapping the church in an amnionic embrace.

I have never liked thunder.

It rained, and then it rained harder. The forest floor gurgled.

Thatch fell down around me, chaff filling the air. The first wave of water-mud-debris surged through the church's entryway, magma-thick, consuming. I searched for higher ground inside the church, anything, a way up and over the quagmire. Magnified by my fear of thunder, in moments what had been a rainstorm was now the Apocalypse.

I eyed the mud cross at the front of the church and the froth surging around my feet. This, then, was my choice. I stacked broken bricks on the neatly made cross and crouched on top of it. Balanced, I watched the floodwaters move across the floor. Rain heaved outside. Rapidly, the poorly made mud bricks liquefied in the rain. They crumbled, were washed away. Before long, little was left. The poles stood at rakish angles, holding up latticing covered in shredded thatch. The bricks were unrecognizable, dissolved into tiny particles of silt and stone and human labor.

I continued crouching on the cross, testament to human belief in God's presence in this forest. I remained immobile, folded up on the mud, crouched on a dissolved cross inside a dissolved church in the middle of a rainstorm.

Later, much later, when the downpour sated itself and the clouds fled north to the Congo, I unbuckled my body from the ground. The carefully stacked mud bricks were gone; the timber poles knifed at the sky, supporting no discernible roof. The trees steamed, misty green, as the sun appeared from behind the clouds.

What is left then, after the storm?

Years after the rainstorm in that church, back in the United States, I stood in the shattered remains of a hospital room, staring at an EKG monitor attached to my mother. I watched as my sister's face collapsed, as my dad did his best to hold himself and us together, as my brother gave himself in to the sorrow. We had not planned for this storm; we had not meant to be caught out unawares in the forest. And standing there, I had no idea that sooner than imaginably possible I would be there again, staring down at my father.

I remember when I found out I was going to Africa for the Peace Corps.

"Zambia?" my mother said.

"Zambia, in Africa?" my father followed.

I stood in front of my parents, breathless, gripping the huge manila envelope that had just arrived in the mail. It was from the Peace Corps, and it detailed my volunteer assignment and country. Zambia.

I had no idea where it was.

The three of us went immediately to the enormous wall map in the kitchen. We hunted for the country. Dad found it tucked in below the Congo, centered in southern Africa.

"I can't imagine it," my mother said.

Frozen over my mother's hospital room watching the EKG on the day she died—that was unimaginable. Stacked in close around the bed, my brother, my sister, my dad, our friends Juli and Dave—leaning, watching, waiting. Juli was my mom's close friend, and her hand rested on my mother's tired arm. Juli said her goodbyes, whispered words, the parting of lifelong friends.

Who had foreseen this?

Fast-forward. Less than two years later, my dad died in his own bed at our farm. We watched him take his last breath and stood there stunned. My aunt Lyn covered her face with her hand, a hand that had the same long fingers as her brother's. Miki, our family friend and a retired nurse, articulated what we could not: "He's gone."

Unimaginable.

Unimaginable is the language we use for today. For climatic changes. For sea levels' rising, for disappearing species, for receding glaciers, for enormous hurricanes and long droughts, for high and low temperatures, for tropical birds' appearing in the Arctic, for permafrost becoming impermanent, for climate refugees begging for asylum because they cannot go home. Carbon emission counts are skyrocketing. Disease vectors are changing. I can rattle this list off by heart because I hear it over and over and over. In the media, in academic journals, over coffee with friends in Reykjavik.

As a scientist studying climate change, I live and breathe this information daily. As a trained geographer, I combine the three main academic disciplines—natural sciences, social sciences, and the humanities—to build a better picture of how one of the visible, tangible, key indicators of climate change, the recession of glaciers, is understood by human societies. In essence, to talk about climate change, I talk about glaciers.

I find, though, that talk of glaciers and climate change is crippling in its enormity. Before huge crowds in overheated auditoriums, during small, one-on-one chats in the field, or in casual conversations with coffee baristas—the talk at times rings panicked, hopeless. The changes reported by climate scientists are unimaginable.

Climate change is like the moon in modern discussion—it is witnessed across the earth from a variety of angles and perspectives; individual and collective eyes have observed it, growing and changing over the years; humans have even knelt and touched its surface. However, because most of us have not been there, haven't touched it, kicked up the dust—even as our daily lives are influenced, shaped, pulled by the moon, bathed in the evening light reflected down—we stand removed. At a distance. The moon is over there, and so we remain over here, where we feel it, but we don't interact with it.

Why have we never gone back? Why haven't we reached through the impossible and built science stations and art galleries and solariums to the stars? Why did we give up after the first try?

Often, when people talk to me about my parents, my parents' deaths collapse into an amorphous story of loss dressed in all the classic words of death. There is no unique moment: death has a homogenizing effect. Instead of the moment when my mom died, or the moment when my dad died, or perhaps all those moments before and after, rather, what I experienced becomes hitched to what all members of human society have experienced or will experience. Death is almost unbearably common, as much as it is, simultaneously, almost unimaginable for each of us as individuals.

I observe this to be the same with climate change.

Climate change is talked about in specialized, isolating language couched in layers and layers of jargon, often in a manner that implies it to be a blanket effect happening to the entire world in the exact same ways. And that blanket effect—it is over there, lunar, at a remove, ordinary. There is no unique moment connecting a person to *a* climate change, to *the* climate change, to this tremendously isolating and homogenizing phenomenon. There is no unique experience nor culminating interpretation or story that moves beyond the jargon and acronyms. There has not been a "right now moment" where we greet the unimaginable, interact with it, and let climate change into our homes and hearts and imaginations.

But it is that unimaginable nature of climate change that serves as a furthering of the disconnect, an excuse for inaction, for turning away. We do not look to the future because we cannot see it.

My parents were diagnosed with terminal cancer long before they died. I never imagined their deaths. I never saw death coming.

I have a children's book from the 1950s. Entitled *All About the Arctic and Antarctic*, it observes that the Arctic is slowly melting. Glaciers are receding, vegetation periods are lengthening, and temperatures are increasing. The book today is over sixty years old.

I am thirty years old. For my entire life, the landscape has been fluctuating because of anthropocentric climatic changes. Anthropocentric. That means us—humanity—our actions and choices and emissions and denials and inabilities to work together. This experience has been every bit ordinary *and* unique; it has been happening both to me right here and to people everywhere. I blink and look out the window again—it is the middle of winter and the Cascade Mountains are bare. Where's the snow?

I've lived in and traveled to some of the more remote spaces on this planet, and everywhere I've gone, climatic changes and conversations of climate change have greeted me off the airplane. The tone of the conversation is often similar: climate change is spoken of like it is a ball out of left field, this far off hit that is coming, inevitable. Like the deaths of parents. Inevitable. Inexorable. Unimaginably foreseeable, but something that will happen farther down the line.

Nonetheless, the hard truth is that climate change is not down over the horizon.

The landscape is changing here and now. While it is inherently the nature of the myriad climate systems and human societies on this planet to shift, to grow and shrink, to evolve and prosper, the rapid rate of change widely experienced across Earth is scientifically and culturally understood to be anthropogenic climate change. Unimaginable, but nonetheless today's current reality, much like this fact: my parents are gone. Both radical realities were long in the making before I could ever imagine such a future.

After they died, I was numb. Immobilized. Lost. I had no guidance and no direction. I knew they were gone, yet I spent hours and weeks willing them back. My entire life changed and my known world was gone.

What does it mean to live through the unimaginable? What is on the other side of that?

I couldn't imagine them dying, but they did, and I kept living. In increments, the business of living a life took over. There were chores to do, animals to feed, glaciers to look after, friends to hug and to hold. In truth, there was no choice but to pick up the pieces, to step out again and again into the world. To re-imagine.

I once sat with my friend Mario Taffera over coffee in the Haven Café in Skagway, Alaska. Years before my parents died, I remember telling him that if I ever lost my family, I would never get out of bed again. He replied: "For a day or two, maybe. But then think about how much you'd miss."

Just because we have not yet envisioned a clear future *with* climate change—a future where the causes and impacts of climatic changes are acknowledged and subsequently addressed by human society at various scales—does not mean we cannot move forward into that future, that we cannot re-imagine a world with widespread climatic changes that step by step we interact with, let into our homes, bend our best brains to puzzling over.

I cannot untangle in my mind the scientific study of climate change and the death of my parents. My whole life,

climate change has been progressing, and I cannot understand realistically what has happened to my family without
stepping back and seeing what is happening to this world.
There are too many parallels, and, at times, there is too much
darkness. They can't be separated. The language and, to some
extent, the experiences for both remain deeply similar. Just
as when I could not imagine my parents' deaths, so do I
now hear us talk of climate change as an event we cannot
look beyond, we cannot imagine, of which we cannot see the
other side. This blindness clouds the reality that we are both
in the midst of and on the other side of climate change. The
unimaginable is happening right now. Our job remains, then,
to begin re-imagining courageously.

    "Dear Madam," that letter began. *"I declare my love for
you."*

Spring, 2010, I'm in Alaska looking for my mother's grave. She didn't die here and was never here and had no ties here, not with this place—this graveyard—but I find myself looking all the same. After she died, this happened in any graveyard I wandered into.

This graveyard is in Talkeetna: the Mt. McKinley Climbers' Memorial. It is easy to find: go through the long-armed single gate and walk a dirt path into the center of the cleared area where the trees encroach, overlooking, monitoring. You'll discover a place with weak sunlight filtering through the poplars, defining muted gray ideas of dapple. There is a plaque to the right, ordering by date and alphabet all the climbers who have died here since the mid-'40s. The plaque has ample space for more names, more tragedies.

We're here for a "photographic opportunity." Ten photographers fan out, eyes scanning. They're looking for color and lines and sentiment and story. The lens is the medium through which they've chosen to view the world. I'm camera-less, task-less, wandering. I brought them here and I am done, for the moment. I walk to the memorial plaque and read the names. They are carved into granite, etched into continuance. I recognize some names—living in Alaska makes one a bit more privy to the deaths of the world's adventurers because so many of them come to Alaska to die.

Out of the corner of my eye I see the photographers crouching, angling, searching. They're standing on graves, the spaces between, the benches. I cringe. I know it's likely there aren't bodies in most of these graves—the bodies remain at high altitude—and the graves are simply here as reminders, like the plaque. But I'm bothered. Graveyard etiquette. There might be bodies here.

I stare at a grave near the plaque that's draped with plastic flowers. A patch of green-brown moss grows on the top of the stone. The leaves have been wiped away. I am intrigued because most of the people who died and are buried here are not from Talkeetna; they are from all over the world and came here just to climb the tallest mountain in North America. I wonder who is caring for this grave.

I look around and see that about half the graves in this yard are spruced up, arrayed with candles and flowers and other tokens from the living world. The graves on the periphery appear untended, abandoned, grown over with lichens and mosses and fallen leaves. Who is making such an effort to clean only half the graves? I walk deeper into the graveyard.

It occurs to me then, standing among the graves, that I wish I had my own mother's grave to tend. I wish I could kneel beside it, bend myself into a position that made my legs numb and my back ache—I wish I knew how to keen. I would force seeds into the fresh ground around her grave and I would monitor them, water them, cast light upon them, and when they grew and proved alive and vibrant I would pick them and take them home and place them on my table and will them not to wilt. I would feel the presence of her and warm my face at the perihelion of her gaze.

I am in a dark mood.

The skyline looks like an EKG. The shadows are elongating. David Estrada, my Colombian co-leader on this

National Geographic Student Expedition, is crouched in the corner of the yard, hunched over his camera. He is shooting and shooting and shooting and I can tell by looking at his body that he is very excited. I go to investigate, trying to shake off my morose thoughts. I am working. My personal life should be packed inside a small box and stored in a garage somewhere—at least for the duration of this trip. I am finding it nearly impossible to participate in the surrounding landscape and not draw parallels to my own recent life experiences.

Due to my extensive work on glaciers in Alaska and the Arctic, I am here on this expedition as a scientist, teaching topics of conservation and natural science, telling stories of Alaska and glaciers and everyday life. David is a stunning photographer and teaches the others the nuances of his craft. Together, we work to help everyone on the expedition find their own story in the spaces and places of Alaska.

David has surrounded a small granite slab rooted in the dirt. Debris and lichen creeping over the marker make what was once clearly a rectangle now more of an oblong afterthought. I lean over him. A shiver runs through my body.

The grave is marked in lowercase letters: "unknown female." There is no indication of a burial; the classical sunken six-foot chunk of earth that foots out from the marker is conspicuously absent. There isn't a mound, either, just poorly tended lawn and this granite marker. Due to the grave's letter orientation both David and I might be standing on this woman's grave—if her body is actually interred here. More than likely we're not. Either way, we don't move. David is tinkering with his camera and speaks of light and two-thirds and angles and shutter speeds while I am immobilized. I am deeply aware of reality, but sometimes wishing overrides reality.

I kneel beside the grave.

My mother is dead, and her ashes are spread at our family farm. Why I am honed in on this grave, this space—and wishing mightily that it belonged to her—I cannot readily say. But regardless, I stand rooted.

A few non-timed seconds roll by and I taste them, savoring. I force my wish into existence. A wail builds down in my sacrum, swirls, and gathers volume. This wail is akin to a hurricane, the hurricane that NOAA just theorized as possible: a category six. This wail will displace cities. It travels, moving through my stomach, tugging bile and advancing to my lungs to vacuum away all available oxygen. It stops my heart with a brutal left detour and then I feel the wail enter my esophagus and move up, higher, seeking release.

I crush it—but only just. The dikes I reinforce inside of me are constructed from the realization that I am working, that I am working for National Geographic, and that now is not the time to lose my shit. I stand abruptly and pound my foot in frustration. I kick at the headstone—and realize in mid-motion what I am doing. I change actions midstream and, in pretext of cleaning the headstone, I scrape my foot across the granite marker, brutally. This is reality.

My foot excavates a ton of moss. David shouts. I am about to explain—we've become friends on this expedition and I know I can trust him. But then I look, and he's gesturing.

The moss I moved reveals nuance. The lettering "unknown female" actually reads "unknqwn female." Now there is no chance. David's camera clicks. This lettering is intentional, which leaves me to consider who would request "unknqwn female" on their headstone, and what kind of statement they are making. My mother would not do this. Whoever this is, reality proves that the connection I'm searching for and trying to make is not present here.

I holler across the graveyard, rounding up the photographers. We're to leave in five minutes. Multitudes of eyes meet mine, real and imagined.

My voice echoes around the trees, coming back to me. It sounds hoarse.

Alaska is considered a refugium. I am enamored with that word, "refugium." It implies a place where organisms can flee to ensure the species' survival. Up here, usually what species flee is ice. Glaciation. They watch their landscape slowly get bulldozed by encroaching rivers of ice and they run like hell. During our last major ice age, as the polar ice sheet grew and grew, so much water was sucked into the making of snow and glacial ice that ocean levels dropped globally approximately three hundred feet and left bare huge swathes of what today we think of as traditional seabed. Flora and fauna and people alike trekked across Siberia, across the exposed Bering Land Bridge, and into Alaska. In northern Yukon, archaeologists have been excavating Bluefish Caves, where radiocarbon dating indicates that humans lived there thousands and thousands of years ago. The north. The refugium.

Today, this refugium is seeing a different trend. Flora and fauna and people are again fleeing to this place, but now they're coming from the south and the east and running from warmer winters and less water and queer seasons and all the manifold effects of climate change. So am I, but I'm also fleeing from a slightly different type of climate change: my personal climate has fractured and splintered and the topography of the landscape is unrecognizable and I find the change is anthropogenic and has a name—grief—and I'm looking for refuge.

I'm sitting on the bank of the Tanana River. I've started smoking again, which means I bought a pack surreptitiously from the front desk yesterday and brooded over it for twenty-four hours before I finally yanked out a smoke. It is eleven in the evening and I've just finished rounds, making sure everyone is nested in their cabins and won't need me until six in the morning. David is off scuffling in the bushes trying to photograph porcupines, and I am alone. It feels good.

This entire trip has been lessons in living piled atop one another. I knew doing any job for National Geographic required my full effort and energy, but I didn't realize it would be this consuming. In addition, my being has become permeated with remembrances, which are beginning to take a toll on my strength. I came to Alaska to avoid having to confront the inching parade of calendar checks approaching the first anniversary of my mother's death. I have always fled back to Alaska after experiencing breakable pain. This landscape protects me. Now, however, it is blithe and careless and my haunting continues.

I inhale deeply. I'm working on my third cigarette. I know I will quit again after this pack—which means I'm not sure if I actually started smoking again or if I am just experiencing a momentary relapse. I have fragmented scarring on my lungs, and this makes smoking decisions hard on my body. I know that in the coming days and weeks I will hack and cough miserably, yet I find myself reaching for another as soon as I've finished my third. Already I can feel my lungs tightening. This feels good.

My mother smoked. I never understood the attraction until my first cigarette in my early twenties. I told her about it months later, when I was back home, both of us perched on the front porch. Coffee in hand, I told her how much I liked the nicotine train that rushed through my ears two puffs in from an evening smoke. It slowed everything down, brought a momentary stillness to the environment. My mother understood completely. By then she hadn't smoked in years. She understood, though, and listened. But then she exacted a promise from me that I would not make a habit of it.

Rarely, she told me, do those openly bad decisions that feel so good now pay off in the long run. Long-term consequences are easy to ignore in the bustle of everyday, short-term life.

I remember jetting over the Cascades of Washington State, burning for a cigarette, peering down at the jagged mountain spine and focusing on the complete lack of snowpack. It was August 2009, the day before graduate school started for me and years and years after my mother miraculously survived a horrific car accident that took her leg and left her in the hospital for months. It was exactly two weeks on the dot after my mother officially finished her cancer treatments. Yet I was flying to Seattle from Montana because Dad called to say Mom wasn't going to make it and I needed to get there—get there, now. At the airport in Washington a friend waited for me, the first person I reached as I systematically dialed each person in my phone starting with *A* to find a ride from the airport to the hospital in Puyallup. Misty Freudenstein was in the *M*s.

I knew what I was flying into, but I didn't think about it on the plane. It was beautiful and clear and Idaho looked enchanting and I stared out the window and considered ice. Can we live without glaciers? Yes. But what kind of life would that be? Would we miss the mothering capacity of our global thermostats?

The sky from the plane was the blue of storybooks and stretched over the wing of the jet and up around the horizon. I aimed my eyes north and knew I would go to Alaska as soon as I could, maybe even the moment this was completed. I had bought a one-way ticket out of Montana for a reason: I would not return, graduate school was scrapped, and as soon as this was over I would never physically move again. If I had been one of those animals standing on the Siberian steppe watching a glacier crush my home and creep towards me, I don't think I would have fled. I would have let the ice envelop me and take me away.

I stub my fifth cigarette out on a large river rock. The Alaskan sky is blazing pink and the sun has dropped just below tree line. It will be back in moments. I stand, stretch, point my face skyward. It is the beginning of a new August and almost a year has passed since that plane journey. I did go back to Montana. My father made me promise, and I attended graduate school and put the pieces back together and landed a dream job with National Geographic, and now I've got a wicked nicotine buzz rattling my brain. I also have a couple more hours of work to do before I hit the sack. I walk back up the bank, push through the willow scrub, and disturb a foraging porcupine. I laugh. David's been out looking for these guys, and we have one right here, just steps from the cabin.

The expedition started in Fairbanks. I was leading a trip through the interior of Alaska with ten students and a revolving door of professionals—all sorts of people interested in climate change and conservation. David taught photography to everyone in the most informal sense. The classroom was in the field, on the glaciers, rafting down silt choked rivers. I was in charge of the science component, lecturing and describing and connecting everything that was photographed, recorded, observed.

David and I had met at a work conference in Vermont at the beginning of the summer where we oriented ourselves to our jobs and roles and emergency numbers. Our first introduction involved my commenting on his stylishly trimmed, short, black beard, playfully mocking his fondness for snappy t-shirts, and us both jumping off a culvert into a creek near Putney, Vermont. He was the kind of guy who stole off into the late-night hours to photograph golden light on maple leaves but always showed up in the morning with an extra cup of coffee in hand. David had an immortal jelly-fish tattooed on his shoulder and could make the most accurate imitation of Chewbacca I'd ever heard.

We were instant friends.

After that weekend in Vermont he'd headed to New York and Iceland and I'd gone to Washington and Alaska. We'd worked well as a team together during the conference and were both confident we would continue that pattern in the high north.

Our schedule for the Alaska trip was loose at best, so upon our arrival in Fairbanks that July, David and I decided that we'd take our expedition to Chena Hot Springs. Alaska has more estimated geothermal energy than any other state in America, but this energy has not yet been tapped in any significant way. Chena Hot Springs is leading the charge, and in 2004, they implemented a closed-loop thermal plant that today powers the entire facility and the surrounding operations. With the students, we toured their facility studying the renewable geothermal energy system and envisioning the hot sulfur-waters gurgling just below the surface.

For me, the best part of the tour was the greenhouse. Chena also experiments with independent food production. A magnificent greenhouse was constructed in 2004 that loops itself into the closed circuit heating system, maintaining average indoor temperatures of 78°F, even when, outside, the winter temperature is -56°F. To grow vegetables in Alaska in winter is akin to making magic. My group lingered in the greenhouse, studying the compact structure. Chena grows, among many smaller vegetables, large bushels of hydroponic tomatoes. They grow nine varieties in Dutch baskets with perlite and drip-irrigation.

They are proving what is possible in the high north, dispelling myths of what can be done with clean energy. This type of off-grid, self-sufficiency farming could revolutionize the harsh realities of living in the Alaskan bush.

I wove myself through the hanging vines, lightly touching the green, serrated tips of the plants. I eyed my photographers as they zoomed in on the bright fruit. We'd only been together two days, but already some strong personalities had emerged. I knew who to keep an eye on, who, besides David, would likely snitch a hydroponic tomato.

I'd been preparing all summer for this trip, running and hiking and ensuring my body was fit and toned. I'd secured all the necessary reservations and permits, scheduled talks and lectures and rough ideas for each day. Climate change was the topic of the moment in Alaska, and everyone wanted to talk to us about it. We met with industry representatives, government officials, nonprofit stewards, all sorts of people who made up the social fabric of life in Alaska. What I could not prepare for, however, were the small reminders of personal loss that appeared everywhere, in everything, even as I tried to avoid and ignore them. In Chena, the smell that permeated the greenhouse froze me in my tracks. *Tomatoes.*

Once, years ago, my family sat densely packed together on the sand like dots on a Go board, watching the Mexican sun set over the Pacific. We'd been moleculistic that day, my brother and sister and Dad and I all ranging out into the ocean or down the beach but always returning to the sunny umbrella and vibrant towel that my mother perched upon. Since the car accident that had taken her leg, my mother had not continued her yearly sun pilgrimage to Mexico. Five years after the accident, we finally persuaded her to go once again.

My mother had sat on her beach towel and spent the day staring at the ocean. She was a small woman with long, dark brown hair, long legs, and golden-brown eyes. She tanned easily—she was a sun nut who loved to wander around our farm in cut-off jeans and a tank top, watering the gardens and checking in with her tomatoes.

I noticed that bit-by-bit she exposed her pale white stump to the sun. Her arms and legs were covered with purpura, which looked like deep red or purple bruises, caused by one of the medications she was taking. Prednisone broke down small cell walls and created trace bleeding under her skin.

About midday my brother and Dad picked up Mom and helped her hobble to the water line. They waded in about knee-deep, and the ocean waves splashed the three of them. I stood behind them and took a picture: my dad, shirtless, tan, holding my mother's left arm; my brother, looking just like my dad on the right side; and then my mother in the middle, one foot buried in water and her other leg just a stump whishing on the top of the surf.

My dad was a tall man, well over six foot two, with a huge head of blond hair turning sandy-gray that matched the enormous Yosemite Sam mustache he always wore. When I was a child, I wasn't sure if my dad actually had a top lip because of what I thought was a large caterpillar resting on his face. He was strong and exuded an enormous presence that filled a room, like his laughter.

A man watched me take the picture, and, after we'd returned Mom to her perch on the towel, where she sat exhilarated and flushed, he walked over. He pointed at the silver bracelets she wore stacked up on her left arm. He spoke in Spanish. We couldn't understand him. He pointed at Mom's arms, then back at his. We didn't understand. He stared at Mom, shook his head, and left.

Later, much later, we sat watching the sun set, and the man appeared again. He carried a small basket under his arm.

We thought he was trying to sell us something. But he was persistent, big smiles, and he repeated the only Spanish word I then recognized: "Sí, Sí, Sí."

We watched him closely as he took Mom's arms into his hands and looked at them. He pointed at her bracelets—ropes of silver, one with leaf engravings down either side with a small turquoise stone in the middle, and then one huge silver bracelet with three large ovals of turquoise offset by eight smaller pieces. He smiled appreciatively at her silver.

He reached into his basket and he pulled out a tomato and a knife. We watched, wary and curious. He let go of her left arm and placed her right in his lap. He sliced the tomato quickly and lined the slices up on her arm, covering the purpura. The thick smell of tomato crowded the humid air. He lined slices up on her left arm, then pulled two small tomatoes out of the basket. He gave them to Mom. He stood, he smiled, he left. Mom, curious, kept the tomatoes on her arms, and, like magic, the next day her purpura was much less visible. In a few days, it was gone.

All the rest of that winter and into the following summer, whenever I was around my mother, I smelled tomatoes. She tried to keep after it, but eventually, she grew too tired. Her skin was so frail—just a slight bump on anything and huge, angry purpura would bloom. Sometimes I'd look her over in the morning just to see how many new bruise-like purpura she'd awoken with. Sleeping, bumping into the mattress—it left her battle-sore. I had wondered then at the function of skin. Our skin protects us to a point. But then when it is so weakened that all it can do is ooze, how do you heal the whole thing?

In Chena Hot Springs, smelling that deep tomato smell, I turned inward. I flared my fingers to hold my grief and tattooed over my eyelids the imprint of tomatoes.

I remember cruising the Internet extensively in the months after my mother's death, mindlessly wandering and killing time. I recall one news website: John Roach, for National Geographic News, wrote, "Astronomers studying a nearby star say they've found the first potentially habitable planet— likely a rocky place with an atmosphere, temperate regions, and crucially, liquid water, considered vital for life as we know it."

Roach goes on to write that this planet, named Gliese 581g, orbits the red dwarf star Gliese 581 every thirty-seven days. We're not new to this "Earth-2" possibility; in 2007 another planet in that same system, Gliese 581c, was announced as habitable until its proximity to its sun baked those prospects. Gliese 581d is also thought to be habitable—once an atmosphere is built. But Gliese 581g—this planet is described as falling in the Goldilocks zone (not too hot, not too cold), and, as Roach writes, it is "roughly three times more massive than Earth [and] tidally locked to its star, which means that one side is perpetually basked in daylight, the other side constantly dark. Aliens, if they exist, are most likely to live along the line between shadow and light, a temperate region known as the terminator."

It took scientists eleven years of gazing at the heavens to find Gliese 581g. Some teammates gave up, but others just kept looking. I understand their drive. The day after my mother died, I ransacked the house. My dad was outside and my siblings were somewhere—I don't remember where. I went through her bathroom, her cupboards; I searched her bedroom and the table by her bed; I looked in her bureau and closet; I went upstairs and searched her yarn baskets. I shook down the bookcase, her desk—I looked and looked. I didn't know exactly what I was searching for but I knew I would comprehend it when I found it. Unexplainable, but I kept searching. I sifted through her jewelry box, her medical journal, her case of note cards.

I understand today that then, I was wild with grief; however, even now I don't know how I didn't find something. After all she went through, her amputation and sickness, the COPD and deteriorating lungs, trying and hoping and åwaiting for a transplant, the earring stuck in her esophagus, pneumonia, breast cancer, chemotherapy, and the pain— phantom leg pain, surgery pain, dying pain, cancer pain, heartache pain, breathing pain, losing-mobility-and-not-be-ing-able-to-do-those-things-that-rejuvenate-you pain, the knitting-needles-are-too-heavy pain, the no-one-under-stands pain; how did she not leave me something? Not even a note? I looked and looked.

Scientists say that we can't see Gliese 581g from Earth with the naked eye. It's in the Libra constellation, but we can't see it. We have to trust it is there. But, if you happened to be standing on Gliese 581g, you could look over your shoulder and see our sun easily.

My mother ran out of air. And she died. Her lungs stopped working, but she couldn't get new ones because she had breast cancer. When she died, I felt then as if I'd never get enough air. The landscape of my life had crumbled, and what I had left was a right arm jangling with her bracelets.

We tried to respect her wishes.

My mother does not have a grave. She has a peach tree, yes, but not a grave. A week after her death we all drove from the farm down to Sumner, Washington, to pick up her ashes. She had been compressed, no longer the size of a person, but rather, the size of a non-person. My brother, my sister, my dad, my aunt—no one remarked on the size difference. It bothered me. I wanted something more substantial, something that had weight and angles and food preferences and a particular affinity for nasturtiums. Instead, we had a rectangle, akin to a cereal box, full of ashes. I cut my eyes to the box the entire trip back to the farm. No one in the car spoke much.

What we did was simple: we got home and we had coffee and we got together and we walked the farm. My dad carried Mom. He sprinkled her in the front garden, by the apple trees, in the fields, in the back gardens, near the windmill, and finally, what was left of Mom he rained into the fresh pit we'd dug earlier for the new peach tree. That was it.

There was so little to scatter, and it took so little time.

In the following weeks, I sucked into myself. I lost my human qualities. My eyes were liquid. They reddened permanently. I knew birds with red eyes could see better underwater, and I wondered if my body was adapting.

And each time I wiped my eyes, my arm would jangle, full of silver. I learned in the days after to polish the silver, to not let it go black.

We leave Chena Hot Springs and David and I take our expedition south to Denali National Park. The students and I hike above Eielson Visitor Center, taking the Alpine Trail a thousand feet skyward to Thorofare Ridge. There, the trail ends, opening out into a series of high tundra plateaus and ridges. The students are panting and we've stopped at the ridge crest to turn around and look down at where we began. Eielson, the visitor center located at Mile 66 off the national park road, is nearly invisible.

Denali National Park rebuilt the visitor center a couple of years ago and reopened it in 2008. The new structure is LEED certified at the platinum level. Eielson is partially buried in a low hillside, with extensive tundra mats blanketing its roof and down decks. Besides the ribbon of road, the eyes don't land on the center unless they're looking for it.

Kim Heacox, our trip's resident National Geographic expert writer and photographer, speaks for a moment about why the national park system is hell-bent on demonstrating new green technologies. Kim suggests that we all hike up to Thorofare Ridge, and we stand close together, listening to him describe this place, this park, this mountain. Kim is a famous storyteller, renowned equally for the photographs he makes and the books he writes.

But even then, after a few moments standing there together, Kim falls silent. He places a hand on my shoulder, and there is much said in that gesture. This place contains multitudes. The landscape speaks in italics, the mountain is shrouded with low clouds, and the tundra melds into the base of Denali, where scores of smaller mountains display only their minor bases draped around like lily pads. This is a place of inordinate proportions. It is hushed here.

The ten of us split up. The students are quiet, unused to the open wildness and sore from the steep hike up. I take two of my students, Hilary and Bailey, and we jog across the tundra plateau. Hilary is tall for a teenager, long and leggy, and she towers over my five-foot-three frame. Bailey's thick blond hair whips behind her as she runs, her camera nestled closely in hand.

In all my years of being out here, I still find incomparable the rejuvenation I feel running on tundra. The thick mat of moss, lichen, and low-bush cranberry makes the first few steps feel like the feet are sinking into cloudy foam, and after I pick up a good head of steam, I feel like I am running on a trampoline. Each stride sends me farther, farther, until I am floating and my body does not graze the land.

Five strides in we jettison our packs in a pile and aim for a distant ridgeline. We are there in seconds. I am exhilarated, and when I look at the students I see they feel the same. This place. Here in this landscape I can poke my fingers into glaciers and summon ice ages. Mountains fall down around us, scree rumbles, the plateaus extend and extend.

I turn three-sixty, sussing out the next point. We see a pile of talus and a large drop-off about a mile in front of us. Without words assent is given, and we three take off, three women streaking across a mat of green. The air is clean, brisk, laden with moisture and sunlight. As I run, my adrenaline heightens and I open wide my mouth, deep-smelling through my mouth. My lungs burn from the cigarettes.

I know that I can spend weeks teaching my students about climate change, about what is at stake. Or, I can take them up here. Here they are given the knowledge in seconds. Here, the musty odor of a planet alive is prevalent. I run faster and become lighter. The weight of a long year drops off. The voices in my head quiet, and all I am is reactive, shooting left and right and jumping, surging, pouring over the landscape.

I feel the clicking before I hear it. I turn my head swiftly to the right, and suddenly there are caribou hammering up, over the drop-off, and onto our ridge. They are running in front of us, gaining ground and widening the distance—but we are all heading to the same place. The syncopated clicks from the tendons that roll around a small bone in each of their feet gather momentum and drum out a live beat, the landscape Morse-coding its existence.

We slow, women and caribou, as we approach our target. David appears from the left. He has taken a circuitous route and we've all ended up in the same place. He has a tail of photographers with him and they are clicking, clicking, cameras whirling, shooting the caribou. The caribou have crested the ridge, and they spread out down on the lower plateau. There are about ten of them, and they mingle and graze. They know we are here so they're nervous, but not too nervous, and their wariness is communicated through flicking ears and snorts and the throwing of eyes. David is vibrating with excitement and I cannot speak.

David and I do not know that we are breaking park rules. We only find out later, from officials, that it is against the rules for us to take students hiking in the park, because we are paid by National Geographic Student Expeditions. In the park's eyes, we are illegal guides. David and I do not know that because of this infraction, today will be our last hike in the park. We don't soak up the moment as one does when they know it is the last. Rather, we stand silently in a state of awe because of the immensity and sureness of this place, humbled to be present for this landscape and glad to have finally found it. The weather turns and a cloud creeps over the ridge, enveloping us in a sudden gray-out. Cloud drops moisten my face. We are blinded.

My mother died on a Sunday, and it was not raining. I believed up until that moment that such sadness should begin with a flood, a catastrophic weather event to wash away all normal reality and signal the end of hope. This didn't happen. Instead, I got onto a plane, flew to Washington State, and walked through the sunshine, and she died. I walked out of the hospital and back into that August light and it didn't feel the same and I couldn't make the colors sharp again.

The last time I had seen my mother was the weekend prior to her death. I was sitting on her bed in the early morning, nervously waiting for my brother and sister to arrive. My siblings were going to escort me on the drive from Washington to Montana and graduate school. We were supposed to leave at nine, but it was nine-thirty, and they hadn't arrived yet. My parents and I sipped coffee and watched out their huge bedroom window.

I wove my fingers into themselves and picked at a loose thread on their bedspread. I had never been to Montana and had allotted myself one day to arrive, move in, and unpack before showing up Monday morning for a week of teacher training. Preparation was never my strong suit.

That morning, sitting on my parents' bed, we talked about many things, but I can't remember what they were. The moment was ordinary in its routine—throughout my life, my brother and sister and I would wander into my parents' room in the morning or evening and beach somewhere on their huge bed and chat.

I want to remember that last, specific time, and I try. I want that moment to be special, to be ingrained as the last time we had together. When Grant and Sarah arrived at quarter to ten, amicably arguing about my sister's inability to speed or rush, our family had another cup of coffee together, spread out on their bed. I do remember that we laughed. I remember this clearly. My sister hit my brother playfully in the shoulder—he, the notorious speeder in the family, she, the notorious slowpoke. All of us, laughing. Mom laughing, her dark hair streaked with gray pulled back into a low pony-tail, face crinkled with sun wrinkles, tan, brown eyes dancing. The only mar on her face was the nasal cannula tucked into her nostrils and stretched over her ears like a permanent grin.

I wish I could remember the details better. That entire last year had been a catalogue of these moments—things I wished I could remember more clearly, things I wished I could tell my mom. I keep looking for signs of her so I can tell her my life, my thoughts, the evidence I've found that

I am surrounded by the deep overlap of the human and more-than-human worlds—that this myth of two separate, different worlds, ours and theirs, isn't true.

But how do I say all of this? My mother's trees are still in the yard, but what will they tell me?

Coming back from Alaska, before I returned for my final year of graduate school, I ate a peach off the peach tree. I looked again through the house. I wished she would haunt me. I wished she had a grave. Evidence that she once was. I need it still: its permanence. The way I need Denali to remember Alaska. The way we need any park to remember the Earth, how it was without us.

SEVERAL YEARS BEFORE MY MOTHER DIED, AND YEARS
before cancer wrapped itself coldly around my father's lungs,
I became an iceberg.

I fell into a deep pool of water on the Denver Glacier
by accident, and, while awaiting rescue, I flirted heavily in
the freezing bone-chill with the idea of giving in, the idea
of letting go and rolling over like an iceberg and becoming a
permanent fixture on the glacier.

Glaciers are malleable, flexible, elastic. Neither hard
nor hollow, they inhabit a deliquescent space somewhere
in between. Areas of glacial ice exist so hard and compact
that there isn't room for even oxygen molecules. There are
sections of ice under such immense pressure that they've
morphed into an ice-water-lava substance akin to molten
plastic, only cold. All this in just one glacier. And there are
tens of thousands of glaciers. Approximately ten percent of
the planet's land surface is covered in ice, is deeply, delight-
fully glaciated.

Each glacier is extraordinary, unique, just like people.
Glaciers continually change, like people. On some glaciers,
the changes occur in a day or a single moment; on others,
the changes only happen within a glacial lifetime. Glaciers
are changelings.

A single glance at the surface of a glacier tells what is happening underneath, below, inside. If the surface is smooth, wide, there aren't any obstacles for the glacier to overcome. But where the ice buckles, where it is rough and jagged, where the surface shows a wide field of crevasses and cuts, then underneath, sub rosa, there has been a disturbance, a stress, a sudden newness to the known topography. The stress might be a leftward or rightward flow around a mountain, a sudden uphill or downhill, a change in earth material, an unsettling rumor or a doctor's diagnosis; it might even be the invasion of another glacier, intruding, overtaking.

Crevasses hint, tell truths on the glacier, that below the surface there is turmoil, things are not all that well; there's been an instant reordering of the topography, a sudden pivot indicating a disturbance, perhaps a death in the family or the leaving of a lover. Luckily, crevasses never cut through the glacier to the core. Crevasses are only on the façade, jagged wrinkles on the outermost surface. Whatever caused the hurt, while perhaps not fully known to us, left the ice raw, exposed.

The scars don't last, though, unlike in people. I've seen glaciers pick up sudden speed, intoxicated on a new direction, a smell, a hope—the excitement of incremental speed slamming the ice into itself, stitching, healing the crevasse. Hollywood shows crevasses yawning open on a whim; rarely, however, during actual crevasse conception, does this happen. To create the crevasses, the glacier wrinkles; long days of constant, opposite pressure and force are needed, a slow snapping inevitable. The cuts on the surface of a glacier take a lot longer to create than to heal.

Glaciers, in truth, are not all that unlike us.

When I fell into that pool of icy water on the Denver Glacier outside Skagway, Alaska, my helmeted head was the only thing visible. The rest of my body was isolated, sinking, freezing. Just one small part of me was left dry, uncommitted to the ice world.

I did fight in that water. I tried to move my limbs, to keep circulation, to keep the numbness at bay. I knew I could not give up, that to embrace my metamorphosis from human to iceberg was a surrender, a white flag, an acquiescence, switching off the ears and the mind. It was a ticket for death. But the temptation of cold is a siren's song. The Arctic is littered with stories of naked bodies, of dead men found unclothed, lured in the last moments before death into the peaceful, easy state of hypnotic cold. The body gets so cold it thinks itself into warmth. And that moment of warmth is an invitation to give in, to strip down, to sweat out the seconds before death.

The place where I submerged was in a large crevasse field about a mile up from the terminus of the Denver Glacier in southeast Alaska. There, ice walls had thrust up out of the level surface of the glacier.

At the time, I was working as a member of Skagway's Search and Rescue Team, both as a medic and as a firefighter. We were out on the ice for search and rescue glacier training, practicing crevasse rescue. Out there, often climbers and hikers and tourists found themselves in dangerous situations that required our search team to mobilize quickly.

All glacial ice has the same parameters, but within those parameters, variations are in the millions. It is the same, in essence, for people. The majority of humans have noses, and the majority of glaciers have snouts. Comparisons between us and them are often fun. Glacial ice creation is a trusty, elegant, step-by-step process that involves immense amounts of pressure and time, not unlike the maturing of a person. I'm drawn by the purposefulness of it all, the steady, deliberate creation of glacial ice.

The vastness of glaciers is created by an amalgamation of the miniscule. Imagine your hand, with fingers spread out wide. The hand is one whole snowflake. The palm is the cen-

ter of the flake, and the fingers are the snowflake's fine pha-
langes, the crystalline antennae. When it snows, the flakes fall
from the sky and land atop one another, like one hand resting
on another hand. In most climates, eventually, the piled-up
snow will melt away with spring. But in places where more
snow falls than melts away, glaciers form.

Glaciers form because, as the flakes pile up, multiple
hands on top of each other, pressure and weight builds from
the bottom of the pile upwards. The snowflakes' furry antennae reach out for each other, grasping, holding, molding
into one another to create a uniform of glacial anonymity.
Air is squeezed out of the open spaces between, and then
the snowflakes themselves begin to shed their antennae, their
fingers dissolving down into the gaps. Soon, all that is left
is the snowflake's center, the palm. This slowly hardens and
compacts even more, transitioning into a tiny round granule
called *firn*. Add more pressure, more more more, and soon
the *firn* weld themselves together into glacial ice. This process, this slow, deliberate creation of glacial ice, can take from
fifty to fifty thousand years, but the ice takes just days to destroy, to whisper away in the dulcet tones of climate change.

To me, the formation of glacial ice is a process one can
count on, a reliable planetary progression, a combination of
elements and time that is much more substantial than anything else on Earth. I am drawn to the idea of frozen ice
rivers carving immense landscapes, mere direction differen-
tiating between mountains and valleys. At times, it is easiest
to look at a map and understand the topography through
the lens of glaciers: the flat, scrubbed interiors of Canada
and central Russia raked by glacial ice pushing south; the
uneven ridgelines along the coast of Iceland the product of
so much ice in the center of the island that it was depressed
into Earth's mantle; the curious shapes of Rhode Island and
Manhattan evidence of glacial bulldozers; the mounding up

of planetary materials into enormous moraines; the sharp, plunging fjords of southeast Alaska, Patagonia, New Zealand, Kamchatka, and Norway.

I believe in glacial permanence, even as it melts under my feet and whispers away. This is ice that shapes backdrops and geography and terrain and self-identities.

My submersion within the glacier was an accident, but it was foreseeable. Not unimaginable. I had my crampons sunk into the ice, my search and rescue teammate Paul Johns on my belay. Paul was a big, burly man who worked as a US Border Guard. He was new to climbing and very new to being on ice.

We were situated at the base of a wall and between the ice wall and us was a pool of blue-black glacial water. No one knew how deep the pool was, but then, none of us had gone down into it.

The type of ice climbing we were doing was called top-roping. Other teammates had set an anchor composed of several ice screws up near the top of the ice wall. The rope triangled from Paul to me through the carabiner up top. Paul was about two-thirds of the way up the wall when he hit rotten ice. Rotten ice has a lot of oxygen inside the tiny individual crystals—it's "airy." It's neither structurally sturdy nor strong, and it cannot hold an ice axe that has a human body weighted on it.

Paul sank his ice axe into the rotten ice, flinched at the ice splatter, and leaned back. His axe cleaved out of the wall and he plummeted.

I saw him fall in slow motion, and instantly realized what was going to happen to me. There was no one at fault here. Accidents happen. Had I been tied into a separate anchor, I would not have been pulled into the pool. Crampons will protect you from most directional pressure—except upward. And Paul was above me. As he fell, I was yanked off my feet and into the sky. He managed to catch his axe into the ice and break his fall, but I'd already attained lift-off.

I twisted in the air and then descended in what normal time must have seemed like seconds, a blink, but in my mind it was slow, gradual glacier time that took long minutes. I felt like I was arcing slowly into the pool. I sank. Cold crept in, flash-freezing. I flailed my legs, my arms, churning to keep my head above water. My lungs ached as icy water wrapped itself tightly around my core, squeezing, isolating.

At times, there appears to be a world outside, off the ice, and a separate, insular world on the ice. To differentiate the two is to believe that sometimes otherworldly events occur on the ice, that the rules of living and existing are different here—if they exist at all. On ice, where water melts and freezes in an hour, where polar bears build castles and the white queen waits, paradoxes and mysteries abound.

Things on glacier work in opposites. The sunnier it is, I've learned, the colder it gets on the ice. Clouds act as a blanket, insulating—when they're gone the ice simply reflects all the sunshine and heat back out into space, leaving the surface frigid and deceptively cold. The warmer the sun is, the colder it is here.

Things move on the ice that, in the real world, normally don't. Huge boulders roll up-glacier, swimming along salmon-style against the ice stream. From a distance, I've seen rocks, stranded in the moraines, pick themselves up and walk across the ice plains. When I hike down to the rocks, they're where they were originally—they haven't actually moved as I just saw them do. I attribute a quarter of this miragism to sublimation, a glacial phenomenon where the ice vaporizes instead of melting. Sublimation looks exactly like waves of heat pouring off of tarmac on a hot day. Looking down-glacier, everything gets wavy and dizzy, and the mirage-like effect distorts distances and causes the ice world to move, shake, dance. It makes the rocks creep.

Bright high-July days bring proof-positive identifications of spirits and bears and apocalyptically approved futures looming mere feet outside your range of vision—just down-glacier. Turn your head to make eye contact with your future: it is gone, dancing off into the other world. Once, I saw a tree stepping root by root through a moraine; my friend Mario walked up out of a crevasse and then back down into it, totally immune to gravity. His bright red jacket flashed red, a blur, and he was gone. More than once, on the ice, I saw people who shouldn't be out there, people I knew for a fact were cozily sipping coffee in their warm kitchens somewhere down south. The ice changes the rules of the possible. The rules of the real.

I think glaciers like to appear mysterious, unreal, otherworldly. It lends them a certain level of respect, like the careful handling of an unknown beast. It is their best defense against man, against climate change, and against cannon-bursts of carbon emissions against their port sides.

The cracks, pings, shudders at odd moments: the ice below your feet is changing and shifting, and there is magic involved in the liquescent movement. Out here, I've seen rivers appear and disappear, discovered bird bones seventy feet down in a crevasse, and watched ice spirits summon sudden summer snowstorms.

Glaciers are not dead. Their demise has been foretold by men in warm, white offices far from here, who foretell that 2050 will see a massive extinction of many individual glaciers, that 2080 might remove all traces of them from the entire planet. But they are still very much alive, shaping, creating, birthing landscapes. They may be isolated, but they are still here, present in both our world and theirs, hanging on even as darkness creeps closer. Their deaths may be foretold, but they absolutely are not certain. Those men in warm offices haven't met the glaciers I have.

Once, hiking down-glacier, still a mile from the terminal toe, I found a tree growing on the ice out in the second lateral moraine. It wasn't supposed to be there. Its roots were wriggling through thin moraine soil and were touching bare ice, twisted and gnarled and grasping small rocks. But it grew, wrapped in around itself, isolated from the forest, tucked in, only four feet tall and fighting it out.

Cold does funny things to people. People get so cold they feel hot and get naked. Cold tends to come in and take the mind away, wandering far off, leaving the lonely body and the frightened, sightless eyes. When I fell into that glacial pool, the cold came and took my mind, took it far, far away.

I wasn't scared. I was glad not to be physically present in a place that was painful. I didn't have a way out of the pool, and I had to broadcast faith out into the world that help would come. I could feel my body slowly morphing, transitioning, becoming glacial. My mind fled to the ice world.

I glimpsed all the other possibilities of life, and they left me gasping.

In my mental haze, I remember seeing vividly an icy river running fast in the heat of a high, summer day. There, I walked my mind-body to the edge of the water, where the hot sand rubbed against the frozen surface ice. The ice was as white as the soul of the sun, grainy and exact and insistent in its disintegration.

I saw both my parents, surrounded by ice, and didn't know what this foretold.

I existed briefly in this other world defined by paradoxes

of heat and cold. Heat bore down from the sky and collided with the frigid ice. I watched the ice rub the sand, dislodge and assimilate each granule. I watched a transfer of existence. The hot sand became cold ice with little fanfare.

The ice world I was fully submerged in as I floated and flailed was different from anything I'd seen before. I wanted to linger and I wanted to flee. I was comforted to be so near my parents. Here in this space, they persisted far into old age and healthy passing.

I floated there for seconds of Earth time—but in glacial time, I lived a lifetime.

My trust in other people was well placed. I could not have saved myself on my own from that glacial pool. I needed help, and I got it. I was pulled violently out of the water by a member of my Search and Rescue Team, Jason Jones. Jason was from Spokane, Washington, and wound up in Skagway, Alaska, looking for adventure and open spaces. He came looking for the more-than-ordinary, and pulling me out of a glacial pool certainly qualified. I'm not sure precisely how Jason did what he did, how he fished the hood of my jacket into his hands and then hauled me to the edge of the pool and back into this world.

He told me later that he'd heard a shout and turned immediately. That he'd let his instincts kick in. That he had done what he was trained to do. Jason ran to the edge of the water, grappled with his ice axe, and used it to fish my hood into his hands and then pull. In all, it took him seconds.

To me, there is no state more helpless than knowing what you are supposed to do, knowing all the steps to make a problem better, but being entirely unable, powerless, to do so. To stand idly by and hope help comes, or to completely give in to what you think the future will be: to lose agency in the face of death or hurt. And even if help comes, it is still so hard to hand off your fate to strangers, to kindle hope in otherness. It is better, perhaps, to trust in your present state of isolation.

During our world's last major ice age, isolation was a conceivable form of survival. In southeast Alaska today, if you look up and gaze at the sharp peaks, you're looking at the few remaining ice-free chunks of land from a time when a third of the world was covered in ice.

In southeast Alaska, the ice sheeted everything, gouging out fjords and razoring mountainsides and sloping valleys, everything but what the local First Nations People, the Tlingits, called *Nunataks*: islands in the ice. *Nunataks* are land ports, jagged, wrinkled peaks of ice-free isolation that stood alone, year after year, for the two million years of the Pleistocene. Hidden up there in bare rock were seeds and plants and mosses and insects and animals, waiting, waiting, and when the ice blew its final breath of frigid power over the world and slowly crept back, they returned, and their isolation was validated.

Just as one glance at a glacial surface can determine its topography, so too can one look at human skin determine the level of cold injury. Simple cold injury shows the skin turning white. Blood has retreated, but not given up. It will return with the correct incentive—gloves, heat, hot breath on cupped hands. A little care, a little consideration. But deeper cold injury shows itself with darker shades—blood has packed up everything and blackly left, with little hope of return.

My cold injury was simple: I was cold and miserable and

numb and burning hot. Jason did all the right things. Got the wet layers off. Got the dry layers on. Made me move my body. Lying in the tent after being submerged in glacier water, I was in agony. White, the pain came in waves, flashing up my body. Jason kept checking in on me and I kept growling at him. I felt like I didn't have skin and the pain was steel wool rubbing over my exposed flesh, grating.

Hours passed. My friends took turns, watching, helping me. They wrapped me up firmly, tucked into three sleeping bags nestled into one another, with me at the core of down. It was hot. I couldn't feel my feet and kept waving my lower legs mermaid-style, the entire pile of legs and blankets jerking up and down spastically.

I felt like a fish. Literally. With a sleeping-bag tailfin.

My mind wasn't working properly, and I remember trying to mentally check through my body to assure myself that all my limbs were still there. I was afraid because I couldn't feel my legs; I assumed they weren't there anymore. I remember thinking that, because I felt like a fish encased in down sleeping bags, I must have fish-like properties.

If things didn't go well, I possibly could swim out of the tent; I could transfigure my new, legless body to survive underwater. I could exist without limbs. I reassured myself that perhaps I could forge out a new life in the waterways of this world.

That night I tried to sleep, but instead, I had waking dreams as the glacier under me cracked and groaned. I could hear pinging, hissing, as air was released from the ice. Bundled up, scared, I browsed my mind. I dreamt of wild futures. Of the unimaginable.

Halfway through the night, the glacial world, the one without the normal world's rules, started to creep in, intrude; the other world receded and shifted away and the ice world became real. I dreamt then of our farm, of my dad's water trough and his fish named Albert. In my glacial state, Albert and I melded, became one.

Albert was my dad's fish. My dad bought him as a small fry, one amid a scoop of koi fish in a plastic sack full of water. Albert arrived originally at our farm with ten other family members. Unfortunately, that winter the cold set in, and the water trough down in the corral where my dad placed his fish family froze over, the ice getting thicker and thicker until it appeared to us that the tank was frozen solid. We never bothered to break up the ice that winter. But come the spring thaw, my dad discovered Albert at the bottom of the trough, his bright orange, two-inch body darting through the flowing water.

I don't know how Albert survived the winter: his demise was assured, his doom foretold, especially as, early on that winter, Dad had started seeing dead fish bodies in the icy slush. We knew the fish couldn't survive such a major predicted alteration to their landscape, their water trough.

I wonder what Albert must have thought, each day waking up to his tank's getting smaller and smaller, darker and darker, the ceiling creeping down. I wonder what he imagined.

Out on the Denver Glacier, frozen, heavily wrapped in down blankets and surrounded by hot beverages and warm bodies, I knew what Albert must have felt. I knew he could not have been scared that winter in his trough. I think, instead, he showed all the signs of having been resourceful, hopeful. Albert believed in his own fish-normality, his simple fish-life. He trusted in it. He might have experienced the worst possible thing for a fish, but he knew he'd survive.

I think Albert must have gathered and assessed his resources, perhaps planned ahead. He found himself a small pocket of water and air in the bottom of the trough, and all winter long he must have glided through lonely, tight laps, humming to himself, counting the days. If Albert's feelings were anything like mine, he must have felt a flash of fish-despair as, one by one, his brothers and sisters and mother and father froze solid. He must have watched, speechless, as their bodies became part of the fate that isolated him—creeping closer and closer.

But as bad as this must have been for him, Albert had to have had some form of fish-hope. As he circled the cold graveyard of his family, he believed in this world's ability to sustain life. He believed in this world's grace. Albert knew eventually a thaw would come. He did not give in; he simply hummed and kept moving his fins and waited. Perhaps he took a moment to consider the idea that, if he moved his fins marginally faster, perhaps he himself could melt the ice. Perhaps he even, courageously, tried it.

Unlike people, glaciers have clearly visible lines of equilibrium. You can look right up the mountainside and see exactly where the ice keeps itself in balance, where the stability and symmetry of the glacier is maintained. Anything above that line of equilibrium is the accumulation zone for the glacier. Everywhere below that line is the ablation zone. All the new snow and ice is received, logged in, processed in the accumulation zone. When processing is complete, the factory pushes the ice into the ablation zone, where it melts, or sublimates, or, as perhaps happened more than once, it is chipped away to adorn a tasty martini.

One glance can tell you what a glacier is doing. The higher the equilibrium line is, the more quickly the glacier will vanish. The less snow or water a glacier is able to take in, the more a glacier will shrink, like a person no longer able to hope or have faith. If the line gets pushed all the way to the mountaintop, the glacier is not long for our world. A healthy glacier shows a good balance, an even keel between input and output. There has to be snowfall—something, going in, feeding the ice, enflaming the glacial soul, energy and stimulation and direction and help, and if that weakens or goes away, the entire being becomes unhealthy and rots from the inside out. After a certain point, it just can't recover.

I think that, at times, we place high value on isolation, and in the subsequent return of life. Isolation implies, somehow, a coming back to life: a rebirth. Something new. When we walk through dark places alone and emerge whole, alive again, this is cause for celebration among those who knew us before.

When I was pulled from the glacial pool, I was cared for, tended by friends. I didn't lose my toes, my legs. The only real trace left over from that experience is that it seems now to take a lot longer to warm up whenever I get cold. But my friends—they talk about how close a call that was, how it could have gone so poorly. They talk of how it happened in seconds. In a blink.

Listening to reports of the pending disappearance of certain glaciers in twenty, thirty years, this takes the mind to gloomy places. When they say there is nothing they can do—unless climate change is slowed down—and the best glacier doctors admit it is done, over, an ensuing helplessness can sneak in. Because somehow, actively challenging climate change doesn't seem to be the viable solution, and somehow, those early reports of glacier death seem inevitable, foreclosed upon. But then I think of Albert, the fish circling his trough through lonely dark winters.

At times, I am like him. I am in a dark place, and it is only getting colder. But then, there is a difference between us, and this difference can be the redemption. What differentiates me from Albert is that I can leave the trough. I can read the science, understand the prescription of climate change; I can work to alter the path of the people of this planet. I know that it does not have to end poorly. We are not stuck on the path down which we've started. If we want to, we can break the bonds that constrict us. We have made our own boundaries.

Today, I can sit on a deck in Alaska and gaze across the Lynn Canal and watch the Harding Glacier slowly creep back, slip up the steep mountainside and curl into itself, alone. That glacier, I know, hasn't given in, hasn't thrown down a white flag of acquiescence. Its doom has been predicted, but there remains a fighting chance. The scary part, however, is that I know this chance, in many ways, hinges on me, on my choices, and on those of my friends, my family, and all people living in this time of widespread climatic changes.

What I know is that for as close as we can come to the brink, the edge, when all the paperwork is signed and things are scientifically declared and the future is planned out, I think the future is still undetermined.

But sometimes we can't return until we begin the process of comprehending and contextualizing the new geography we find ourselves in, the raw, new landscape carved overnight by unimagined forces. I'm more and more aware that, as the ice recedes, this world we live in becomes more unlivable for humans. People need glaciers, just as glaciers now need us. Sudden crevasses in our lives can leave us helpless and alone, but we are never isolated for long. What makes up a glacier, I remember, is millions and millions of little snowflakes, reaching out to one another, grasping hands.

SHAUNA THOMAS AND I CLIMBED TO UPPER DEWEY LAKE above Skagway. It was the summer after my mother's death, and we started around seven in the evening, meeting in the middle of town. It was clear; the sky, bright and cool and calm. I carried the tent, Shauna, the stove. The decision to go was made on a whim, during a quick discussion earlier that afternoon. Originally we kicked around the idea of just hiking up and back, but I pushed for us to camp. I needed the peace of Upper.

Skagway, Alaska, sits at sea level, and Upper is notched at 3,097 feet. Each step on the hike up is an intimate interaction with acute elevation gain. It is so easy on the steep hike to turn around, to retreat, to give in. We made it in two hours and forty minutes.

Upper is a small body of water nestled in a bowled-out sub-alpine basin. The trail begins with a series of sharp switchbacks, levels off after fifteen minutes, loops Lower Dewey Lake, and then aims skyward with razored slants in the mountain the rest of the way up. At the top are a cabin and fall-off views of Skagway, the Lynn Canal, the Chilkats, glaciers and more glaciers, the world, my world. This is the place where, over the years, I have garnered the courage to step off the proverbial edge and commit myself to people, ideas, organizations, schools, identities. To hope.

Shauna and I didn't stay in the cabin; instead, we pitched the tent near it and used the cabin's fire pit. The cabin was empty, renter-less—we had the rim of the world to ourselves. There is an implicit joy in sharing knowledge and confidence of the surrounding environment with a woman I consider a soul friend.

Shauna and I have known each other for over ten years. She and her husband, John, live year-round in Skagway with their two daughters. I initially met Shauna my second summer in town, when she was pregnant with her second daughter, Piper. We lived a couple of rooms down from each other in the Westmark Hotel. Half the hotel was actually a hotel while the other half was employee housing for several different companies. I drove a bus for tourists that summer; Shauna worked as a guide at the local Jewell Gardens. We'd meet most evenings in the tiny kitchen that the entire floor of the hotel shared. We connected, and over the years she has been a colorful anchor to whom I return. She's five foot one, about a hundred pounds, spry, smart, kind, with a habit of regularly cutting, dyeing, or otherwise dramatically changing her hairstyle with two seconds' notice. An avid mountain biker, when she and John married at Lower Lake the previous year, they biked the switchbacks down the mountain to celebrate.

Back at Upper, Shauna went off to break low-lying dry branches from the subalpine firs and spruce; I popped pitch blisters on the firs, oozing the sap onto a twig wrapped with black old man's beard lichen. Instant torch, highly flammable.

We met at the fire pit. I lit my torch, she teepee-ed the dry sticks—our fire raged.

There are things I want to say to Shauna. As I approach the first anniversary of my mom's death, the bigger questions are beginning to loom. I don't know what I am supposed to do with myself. She died the day before graduate school started. I began and completed that school year, but I can't remember a thing I learned. I don't know where the last year went. And now, what is ahead is getting darker.

Staring into that fire, I felt like I was beginning to fade. I could feel it: an ache in my bones, a restlessness in my dreams, a wandering of my thoughts. I'd regularly float in and out of myself. Two days prior to the hike I was sitting on a deck in town, basking in the sun, enjoying the company of good friends—and then I mentally checked out, packed my bags, and left.

I had lived another lifetime and yet just seconds had passed on the deck. I had let my mind drift, and I'd left and lived a life in which I hadn't watched my mother die. I had lived a life in which she grew old, peacefully, in which we talked to one another and she guided me, helped me, supported me. In which she was not sick, and she and Dad worked their farm and lived their vision, and I raised a family and brought my children to them for summer stays, and coffee breaks took all day and there wasn't an oxygen cord choking my mother's movements. I lived this life, full of the promise of my mother's wise hand; I tasted it; I believed it was there to be found in the brightness of an Alaskan sun. In this parallel life, I swam in the audacity of family.

But then I returned, and opened my eyes, aged and hopeful, and I discovered that nothing had changed. The disorientation was crushing, inducing immediate seasickness as my realities swayed and threatened to capsize. I fled the deck, seeking solace in the cool basement, where I curled myself up into a cat-sized ball, tucked a safety tail around my limbs, and dropped my eyelids, vault doors locked closed—impenetrable. I retold myself the other story, the one in which it was different, not like this. I repeated this story over and over until my breathing normalized, my heart calmed, and I drifted into a shallow sleep.

When my mother died, I did three things. The first: I delayed one week before I started graduate school, working on a master of science at the University of Montana. The second: I began calling my father every night. Previously, we'd spoken once a week. This changed immediately. And the third: I lay each day on the lightly carpeted cement floor of my garage-turned-studio-apartment in Missoula, Montana, and told myself stories. Eyes closed, ankles crossed, I told myself stories from the hundreds of books I'd read, stories from my own life reshaped, re-imagined, stories of the impossible and the grand. I fantasized, in detail, about leaving this planet, about building a new home on a new world that took lifetimes of chemically induced sleep to reach.

Many of the stories I told myself after my mother died I have since lost—but the detail, and the image, of that new home on a new world remains with me, vivid, real.

I grew skilled at telling myself stories. If I dreamed myself in the ocean, swimming, basking in the tropical sun—I felt in my body the gentle rocking of the waves. My face would warm in the glow of the imagined sun. Once, I even tasted salt upon my lips.

I passed hours lying on the floor. But then, a noise outside—a car racing up the alley, my neighbors yelling, a slamming door, my phone ringing—something would intrude and I would open my eyes and be back in the dark room, alone, bone-cold from lying on the thinly carpeted cement. Getting up and turning on the lights just pushed the gloom to the corners of the wood-paneled space and informed me in cool Montana tones that my mother was still dead. I consistently fought the urge to lie back, to tell myself more stories.

During the days, when I returned from classes at the university and came pushing through the heavy door, rattling the ring of bells I'd hung around the knob, the first thing I did was glance at that space on the floor, midway between the refrigerator and the gas heater. Like most people, coming home and instantly flipping on a television, so too did I fight the habit of coming home and immediately distracting myself. But I relied on my mind because I didn't own a real television. Nothing else sufficed. I wanted to walk into my apartment and flip on a mental story. I didn't feel like talking, and I couldn't read a book. Reading was a bond between my mother and me, a language we both understood and loved and exchanged back and forth, and reading felt hollow without her.

My earliest memories of my mother are the sounds of her reading. She, in the rocking chair, creaking, crunching on ice cubes, shuffling newspaper pages right outside my bedroom door. She'd sit for hours going through the paper. I, a child, awake long after bed time, lying in my bed inside my bedroom and pretending to read the paper, too. I'd make up what she was reading, what stories held her interest, and which ones made her chuckle. Her sounds were imprinted in my mind: the absentminded exhalation of smoke from her cigarette like the release of water from a tide pool, her bracelets sounding discord as they glanced off the wooden arm of the rocking chair. Legs crossed, one leg firmly on the floor while the other bounced absentmindedly. The light shift of the rocking chair's leather seat as she shifted positions. Motherly noises.

When I grew older, I remember sitting on the brown carpet at her feet after she'd come home from work. Sometimes, when money was tight, my mother would work up in Elbe, Washington, waitressing and tending bar into the late hours. She'd come home tired and sink into her chair, and if Dad wasn't home, my brother or sister or I would pull off her knee-high, brown leather boots and listen to her sigh contentedly. Home and foot-free. We'd hand her a drink of water, and the black footstool by her chair always held the day's newspaper—heavily rifled through by all of us, but the day's news nonetheless.

My mother always read, but she didn't pick up real reading speed until I was a teenager. Midway through my teens, I worked an evening shift at a restaurant on the Puyallup Indian Reservation in Washington State. After my shift ended at eleven, but before I made my hour-long drive home back up the mountain, I'd often stop at a Safeway and wander.

It was a way for me to wind down before the drive, a time just for myself, when I could meander the orderly rows and investigate all the groceries. We didn't have much extra money growing up, and somehow, inexplicably, looking at all the things I might buy contented me.

Even today, wandering the aisles soothes me.

One time, I walked into the library section of Safeway, a whole row of magazines and coloring books and paperbacks. I bought a book; I can't remember which one, but I think it was an Oprah's Book Club book with the huge "O" splashed across the cover.

I remember driving back up the mountain to the farm and sprawling out across my mother's bed where she was sitting up waiting for me. She always waited, keeping the light on. At that time, Dad was working long hours in Alaska and commuting back and forth from the farm. At night, my mother waited up for me to come home, her lonely reading light shining like a beacon.

"I brought this home for you," I remember telling her, holding out the book from Safeway.

"Thank you. That was thoughtful."

"I haven't read it. But I'll read it after you?"

"I'd like that."

I handed her the book. This was a year or two after she'd lost her leg, so now, instead of perching in her rocking chair, which rubbed her stump funny, she'd sit up in bed late into the night and shuffle through the newspaper. I remember giving her the book and then, as days and days went by, and she read the book over and over, I remember her asking for another, and another, and I kept bringing them.

Here, finally, was a good task for me. At first, I think the books were all strong female character novels, entertaining and popular, the type I could find at a Safeway grocery store. But then we branched out, going into adventure, mystery, memoir—she opened an account on Amazon. I started reading more and more nature writing, ecology, place-based stories. We exchanged, piled up more and more books. It was hard to tell which books were hers, which were mine. It didn't really matter.

With her gone—blink—I couldn't read anymore. Journal articles and reports stacked up to be read for school, and they were a chore I got through laboriously. The books I reveled in, the ones that brought me joy, the ones we would talk about and reference for years—those stood quiet, dusty, unused. I couldn't look at them.

For years, separated by geography and jobs and intention, we bridged those gaps by reading and sharing books. Sometimes, what she couldn't say to me explicitly she would say by handing me a specific book. We worked on our mother-daughter relationship by talking about stories. She mothered me through literature. With her gone, I slipped into a morass, floated down, and told myself stories. I didn't see another way out. I called my dad, I showed up to classes, and, every other moment, I retreated into my mind. She did not leave me instructions, directions for joy. No note, no letter: she left me no guidance when she died.

Joan Didion writes in her book *The Year of Magical Thinking* that after her husband died, "I did not yet have the concentration to work but I could straighten my house, I could get on top of things, I could deal with my unopened mail. That I was only now beginning the process of mourning did not occur to me. Until now I had been able only to grieve, not mourn. Grief was passive. Grief happened. Mourning, the act of dealing with grief, required attention."

I read Didion in hopes of finding my mother in the pages of her book. There must be proof that the floundering I experienced in the wake of her death was not singular. Gratefully, I recognized in myself the signs Didion wrote of, the impasse of grief and sadness. Today, I understand grief as a sensation of heaviness, a grain of sand stuck in the eye, the waning of hours deep into the night watching shadows play across the ceiling. Grief is powerless, this thing unimaginable, a tectonic smack, the desolation of having the birds quiet down when you walk the forest path. It is leaving the milk on the counter for a week; it is not looking up when crossing the street; it is cold. It is, in essence, unsurvivable.

Mourning is the digesting of food, the early light, the red panties under all that black, the pouring of jasmine tea in an unlit room, the awareness of soil. Mourning is continuous and endless, a birthing of emotion in a closed brain. It is the willingness to pour words from mouth, the automatic reaching across the bed; it is buying a gift for someone who is dead. It is an act of surviving.

Didion does not say when the grieving stops and the mourning begins. She does not write of finding clear, definite instructions lying about, nor of the recognition within herself of the proper instinctual response to grief. What shifts to suddenly allow color into the room? What changes to suddenly point a way forward, to create a future?

Perhaps, at some point, the finality of loss stops being final. Perhaps there is that unseen, unimagined moment when stewing in bed, when turning inward, when giving in to a future uncertain, when all that becomes unbearable for an individual and a people. When that actual moment occurs is entirely unknowable, but when it does, that is the moment to be seized. There is strength in getting out of bed and participating again in life. Immobilization can be a state of existence, but it can't be a permanent state. It is simply too easy to give in, to luxuriate in the indolent, dark waves of grief. Once done, no choices are expected, no action required; all responsibility is negated. Giving in, then, becomes the ultimate non-choice. But then, so much—everything—becomes lost. It is, undoubtedly, better to bank on that single moment when enough is enough, and forward motion, movement, sunlight, getting up, reaching forward, the making of choices and the ability to be wrong and right, the eating of a good sandwich in the presence of friends, are exactly what is sought.

I lived six blocks from the university the months after my mother died, roughly a thirteen-minute walk. The route from home to school involved one left turn out of my alley and a right turn onto campus. In the month following her death, I became lost numerous times, in one instance ending up at an Albertson's grocery store thirty minutes from my house and in the opposite direction. I never told anyone because I was ashamed of my complete inability to account for my time, my loss of direction. I'd leave my house and put my feet in motion and be surprised at where I ended up. Once, I failed to show up for a class I taught because when I came to, I was, somehow, standing by the river that runs through Missoula forty minutes past the hour.

At some point, a few months after she died, I started

trying to counter my inexplicable habit of wandering. I did not accept my fate as permanent. I knew I could change. I left my house earlier; I endeavored to pay attention. I tried to become absorbed in my present state so much that I wouldn't let it float on without me. I counted and identified lichen varieties growing on the cement and rocks and trees and fences. After a while, it worked. I stopped getting lost. My actions required Didion's style of attention.

There is an isolating simplicity in grief. The ways we relate to the world get much more basic—childish. Paying attention was too complex for me, so, unnoticed, I stopped. After the initial days, I responded only to instincts. I ate and slept and ate and told myself stories, and I kept my radar out for someone to tell me what to do.

I needed instructions.

I was a whole person who disintegrated when my mother died. My grief splintered into rain and ran down the gutters and into the rivers and oceans and seeded clouds that raced back to shed on a dreary day. I assumed a fluid form, with no direction, that floated along, passive, until enough was enough and I was reminded to re-imagine, rebuild, re-coup. Somehow, I midwifed seeds of instinct and hope and direction that grew into a new person.

I remember vividly a conversation I had with David Estrada, my co-leader from my first National Geographic expedition. It was when we first met, and he had his shirtsleeves rolled up over his shoulders. He was wearing a blue, cotton t-shirt that looked like it enjoyed being stretched up around his neck. We were standing together in a sheep field in Vermont, overlooking the tents we had camped in the night before. We'd met for the first time the day before to plan out the logistics of our trip to Alaska.

"Is there a man in there?" I asked him, gesturing at the large, squirmy tattoo on his arm.

"Ya. Wrapped up in the tentacles."

"That is really cool."

David grins. "Ya. And it's immortal, the jellyfish. It's called, wait; let me remember. Ah, *Turritopsis dohrnii.*"

We continued discussing the blue tattoo on his upper arm and shoulder. It depicted a large jellyfish spread out over his biceps, ominously encompassing the outline of a human figure.

"Did you know," I said, kicking my nerd self into high gear, "that the immortal jellyfish is considered an invasive species? It gets sucked up into ships' bilge pumps and spread around the world."

"I like that they can live forever."

I nodded in agreement. The jellyfish is incredible. It grows up, or grows down, at will. When its survival is threatened, such as by food scarcity or trauma, "instead of sure death, [*Turritopsis*] transforms all of its existing cells into a younger state," writes Pennsylvania State University researcher Maria Pia Miglietta.

In the event of trauma, the fully developed, well-educated, and successful jellyfish will turn inward, slowly transmogrifying into an unrecognizable pile of cells. When the stress or trauma has passed, instinctually the cells reassemble into infant jellyfish forms called polyps that scream and holler and wave miniature baby rattles.

Magic happens when the polyps grow up.

What was once a muscle cell in the mother jellyfish transforms into nerve endings or completely different cells in the new individual. The small, perfect nose of the former mother might be the ring finger, or ring tentacle, in the new, regrown jellyfish. While genetically identical, the new mother has scrambled cells, reordered into new jobs and tasks from the previous jellyfish. In essence, she is the same person, but her landscape has changed, and she's different.

David rolled his shirt back down his arm.

"Do you have any tattoos?" he wanted to know.

I nodded. Of course. I'd never had the luxury of growing up or down as circumstances required, like a jellyfish, but repeatedly, I have experienced massive changes to my own environment, shiftings of the landscape, standing on the edge of uncertain futures. Wandering the streets in Missoula after my mother's death, I felt unmoored, almost transmogrified into something unrecognizable. Often, I didn't know what to do. There were no instructions. So I tattooed a small quasi-roadmap onto my upper shoulder, lines and lines and lines running as a reminder of the multitudes of paths we can all take to survive and perhaps end up at the same blue dot. The words underneath the map are my mom's, written to me, quoted, I think, from one of the books we shared.

Upper Lake is the kind of place where you can bare your soul for no other reason than because everything else up there is bare. The trees are lean, twisted, short, bent with the reminder of winter snow. The lake is clear, turquoise, packed with minuscule glacial silt. Across the lake, the peaks of the mountains soar skyward. Turn all around, and the landscape is pierced with glacial peaks: the topography is raked with a jagged comb. They all look climbable, accessible from this vantage. It is only another 2,313 feet to top out on either of the Twin Dewey Peaks above Upper.

Distances here are misleading, though. That last two thousand feet are rimmed with scree and patchworks of snow and rock. Peaking out requires equipment, instructions, safety nets, partners, wings. I am acutely aware that I lack all these.

The rocks surrounding the lake are gray and white and mutely adorned with crustose lichen. There is little color here. I jostle around for a while, greeting familiar rocks and eyeing the lichen, looking for unknowns. There are so many lichens I don't know the names for; in a pinch, I'll make up a name. I don't stray far, because I've been waiting for this time with Shauna for months now—a spare moment in our friendship treadmill where we can step outside our norm and shift the dynamic. Up here, the pressures of work and families and friends don't really have traction. It has been ten months now since my mom's death. I am hoping that my friend, this courageous, open-minded mother of two, will lift her wing just an inch and take me in.

Shauna heats the pad thai and I ready our cups for tea. I'm addicted to Montana Tea and Spice Company's Night on Glacier Bay tea. It's smooth and flavorful and spicy and licorice-y and I breathe it more than I sip it. I drop a bag into her mug and one into mine, and I wait. I've tended the fire, we've erected the tent, sleeping bags are in, the ground mats are inflated. I've pulled on another sweater; it is after eleven, pushing midnight, and the sky is still bright. However, a chill has moved in to remind us that, though it is light, it is still night.

I huddle in closer to the fire. Shauna brings the noodles over and we eat companionably. She's telling stories about how she and her husband, John, first met. I'm exchanging stories from the last year of my life—but my mind is not on our stories.

What I want to say is simple: I can't do this without my mother. Not the sitting here talking to Shauna—that's fine. But what I can't do is this business of living. It is not supposed to be a solitary act. I am told, time and again, that time heals. That as the weeks, months, years pass, I will become more and more okay. I will find guidance from other friends and family, I will see evidence of my mother in myself, and that will comfort me. But up here, sitting around the fire, I don't want to go back down to that life waiting for me in the valley.

I am melted down, washed away, and I am struggling now to stay in one form, as one person. It all feels like too much. How do I conceptualize what has happened, what I could not imagine, and what will happen in the future? Previously I had ideas of my future, and very quickly, they changed. I had done nothing to cause all this, but somehow my mother died, and now, sitting here at the fire, I tell Shauna that my father is also ill, and that he is predicted not to make it either. What I trusted in failed. And now, with so much uncertain, I do not know what to do. Whom do I turn to for guidance? I am not getting more and more okay. Instead, I think about my mother constantly. I miss her. I dream about her, I talk to her, I ache, I carry her up mountains, I look to her for advice she cannot articulate.

I look for signs from her in the clouds, in the lichen formations, in the smells of any room I enter. I ask for her nightly before I sleep. I laugh and hope she hears. I look for her everywhere. I need her to tell me what to do, how to care for my dad.

I feel like I have splintered. This is not getting better.

I have too many questions and I don't know how to phrase them in a way that does not flood Shauna. She talks about her hopes for her girls. She gestures and the Alaskan sky is bright behind her. The mountains remain in silhouette.

I am awash with a fierceness, and I lean over the fire to tell Shauna all these things in my heart, but instead I tell her never to leave her girls. But the words don't come out like I meant; rather, they are soft and weak and garbled towards the end. I look into the bright sky and remember that, throughout human history, people have found guidance in the night sky. I strain my eyes for the stars I know are there, but it is simply too bright to see them.

I'm not alone in fading, in melting, in feeling like my soul was washing through and out of my body. Late nights endlessly scouring the Internet produce something interesting: scientists from Norway and Finland discovered that during the harshness of cold winters, to conserve energy and ensure survival, brown trout physically shrink themselves. Called "over-winter anorexia" by researchers, it appears that, as winter sets in, the fish decrease their appetite. It is as if they simply hunker down and will their bodies to shrink in size. The jelly-like substance within the vertebral discs of the fishes' spinal columns compresses, shortening the fishes' overall body lengths. The actual process of how the fish do this isn't fully known. But Dr. Huusko reported to the BBC News that "apparent triggers for shrinkage were food and feeding stress in connection with environmental conditions generally poor for growth and survival."

Somehow, in the transition from summer to winter, from light to dark, these fish make an assessment of their landscape. Collectively, they group together, leaderless. They suss out their food stores, the likelihood of nutrition and variety. They crowd together and exchange news about the weather, the pending winter forecast. They do the calculations. They read the scientific reports on climate change. They attempt to imagine. To predict the future.

I know I should be talking to my friends, reading self-help books. But again and again I find comparisons between my present state and the surprises of nature. I can put myself in the place of those damned fish and know that they each paused and wondered, "What should I do to survive? What have I done so far? What am I willing to lose?"

Is it instinct that demands such a brutal measure? Or the ties of peers? Of family? Where is the mother fish, gently swaying among the masses, reminding them that fins are necessary, that no one is asking them to cope without bellies or opalescent scales. That these times of darkness will pass.

Everything takes a big step back and waits, pauses, holds its breath, and strains to see in the dark. Scanning the headlines in the human world shows us a planet reacting differently than ever before. Bigger storms, bigger fires, higher temperatures. Known landscapes are breaking down. And we sit; we wait. We hold our breath. What can we do to survive?

My brother repeats to me, over and over, what he learned from Mom. In times of stress it is best to wait three days before acting. Three days. Just wait.

When do you know the time for waiting is over? When is enough simply enough?

I think the brown trout have secrets. They know things we don't. They don't find a cold finger of oxygenated water whisking through the river ice and set up camp for winter. They don't pack faith into top-of-the-line backpacks or pad their sleep with feathers. Rather, they take a good, cold look at the lay of the land and make the changes demanded by instinct to ensure their survival. To them, this means removing the spaces between their very vertebrae.

In a way, my mother's body could have related to those fish. To survive, she was compelled to agree to the removal of parts of her body. Her leg was so mangled after the car accident that she wouldn't have been able to walk on it. Doctors removed it. When she discovered she had breast cancer, the best choice for survival was the amputation of her breast.

In our eyes, she was whole. But to survive meant that

breasts and legs became excessive, unnecessary for living. She needed to shed them to live. And after she died, bits and pieces of her remained in the bedroom. Two baby-blue containers holding her prosthetic breasts sat on the bureau. A teetering pile of prosthetic legs leaned in the corner. I didn't know what to do with them after she died.

What was excessive to me? Was keeping her prosthetic pieces necessary for my survival?

Shauna and I sleep late through the morning until the heat of the Alaskan sun drives us from the tent. Upper Lake shines. Clouds puff quickly across the blue sky. It is perfect. We pack rapidly, deconstructing the tent and campsite in minutes. Shauna scans for garbage, while I root around for a snack in the top hatch of my pack. We eat, and then, leaving questions and thoughts and words and hopes unopened along the alpine ridge, we descend.

Always, hiking downhill switchbacks makes me contemplative. This time, a line from a Hari Kunzru novel floats through my mind: "All the world is in the past."

I walk along a path I've already been on, reminding myself of thoughts I had when I came up this trail. The experience is entirely different, however, because I can't lean forward and use my hands to pull myself up steep bits or flail them out for balance. Rather, I slide each foot down, poke around until I find solid footing, then put weight on the leg. And repeat. Countless times I miscalculate and plow dirt with my face. It is painful: I rushed this path on the way up, and now, reversing, I am wallowing. The significance is not lost on me.

To slow my progress and focus my thoughts, I begin lichen hunting. I do this everywhere. At heart, I am a lichen nerd. I have multiple favorites, depending on the location, and out here in the Tongass National Forest, I am quite partial to fairy barf lichen. This blatant favoritism has little to do with its moniker; rather, I am enamored with the way it looks. The lichen grows on damp, spongy, rotting logs and carpets out in large, whitish-green swaths. Dotted—er, barfed—across that green carpet are pale pink circles: spores. Sometimes the spores are bluish; other times, a vibrant pink. It is rare to see this lichen, but it is even rarer to see a green and hot-pink combination in the Tongass. The most common, reliable place I know to see fairy barf around Skagway is on the trail to Icy Lake. I've never seen it on the trail to Upper.

I'm attracted to fairy barf because it reminds me of the first Jackson Pollock I ever saw in person. Pollock's *Blue Poles Number 11, 1952* hung in Melbourne, Australia, several years ago, and I sat with it for hours.

I've been a painter since my teens, a habit that complements my scientific leanings. Vibrant works invariably excite me. Pollock's painting threw me into fits. I remember being on the phone to my mother, bridging the time and geography between us by talking and talking about this painting. I convinced her that Pollock was a painter-ecologist. How could someone look at his splattered works and fail to see the reflection of this small, easily-missed-unless-you're-looking-for-it lichen? She believed me.

Fairy barf is not solid green; rather, it has multitudes of greens stacked upon one another to reflect uniformity. Get down at nose level, and peer into the expanse. The edges are undefined, poised; it will creep or collapse at will. The pink spores are flotations upon a sea of velvet. Run a finger across the top, an even stroke, and feel the chord of recognition sing inside. An electrical charge.

It reminds me of both that trans-Pacific conversation with my mother from years ago and different conversations we had in person. We used to talk about all the diverse plants growing thickly on the farm.

"Emily," she'd tell me, "always keep an eye on that asshole hawthorn tree."

"Mom, a tree can't be an asshole."

"That one is. Watch it."

"Why does it smell so bad?"

"It chooses to woo insects that are attracted to rotting smells like flies, instead of sweet smells."

We both gazed at the hawthorn. It was huge, and small volunteer hawthorns had sprung up all around it. It grew like a weed. It was shaggy and covered in lichen.

"It won't let any other plants get near it. Look at my roses."

We looked at the roses growing along the fence. They grew nice and thick until right up next to the hawthorn, where suddenly, they grew spare and thin.

"The lichen is pretty, at least," I told her. She agreed. We both were drawn to the small plants.

Lichen, at its most base, is simply a term that describes a relationship between algae, fungi, and/or cyanobacteria. That relationship, depending on whom you ask and what lichen you are asking about, can be a parasitic, mutualistic, or symbiotic family. Sometimes, the combination of two or three organisms living together can be mutually beneficial to all parties, while at other times it can be beneficial to one member and not to the others. The fungus side of the family provides the home, the structure, the parameters. The algae and cyanobacteria provide food through sun-farming, or photosynthesis. Fungi found growing in lichen families often aren't found independently, alone, singly; they are vulnerable in the wild. The fungi need all the family members to be working together to survive. Lichen that has become damaged by pollution can't prosper, grow, persist.

Fish from Norway. Immortal jellyfish. Lichen. I look out my window and see so much evidence of compatriots in the more-than-human world. To make sense of the here and now, it is best to look out there. Another example: there is a bee on this planet that doesn't have family. It is solitary, alone, floating along among the planet's fickle winds.

Osmia avosetta has long given rise to myths, but it was only in the late spring of 2010 that disparate research teams discovered the same thing: the lone lady bee constructs her nest carefully, delicately, intentionally, of multi-hued and multi-textured flower petals. Reds, greens, purples, blues, and yellows are all cautiously bitten from the mother flowers, flown to a minor indentation in the ground, and woven, basket-style, into a small, open-ended vessel. A fine plaster of mud is layered on, then another line of flower petals. Provisions in the form of pollen and water are positioned in the bottom of the nest, and then, most precious, the egg is placed, gently, into the nest. A plug of light petals and mud is used to close the nest. With the final act of cinching up the nest, mother bee flies off into the unknown, never to see her offspring.

When the egg hatches, a young bee squeegees into the petal-blessed light and eats the food left by its loving but absent parent. She'll instinctually spin a cocoon around herself and fall into a pollen-induced slumber-slash-daze for ten long, dreamless months, waiting for spring. And when she awakens, the baby bee leaves, flying away on planetary winds from the nest, the only gift from her unknown mother.

Unsurprisingly, this little bee will instinctively seek out colorful petals, drawn to them by the last vestiges of maternal impulse. Unguided, this little bee enacts the same dramas as her mother, carefully creating blue and red and purple and yellow baskets in which to nestle her own young.

I climb down the mountain from Upper with Shauna, heading back into Skagway. I follow in the footsteps of a woman who is not my mother, who is not brown-haired and oxygen-bound. Instead, she is cloaked in bright colors, yellow and red and blue, and I am inexplicably drawn to her. She does not speak in pollen tones, but her light and shine sear through my fog of sadness. We step over the same mud, we marvel over the same lichen, we speculate together on the futures of young girls.

I can't live without my mother, but then, I am not sure the world is asking this of me. I know that in this time of color, this time where grief transmogrifies into mourning, I am paying Didion-like attention to direction. I am surrounded by secret painter-ecologist lichens, strong women, and jellyfish that grow up and down. David keeps a tattoo on his arm to remind himself of the tangled relationships between man and nature; I tend a roadmap tattoo that keeps me moving in the same general direction my parents favored. There will be guidance in all this, I hope. If a solitary mother bee can impart her love for red flowers into the babies she will never see, I have faith that there are instructions for my world and me somewhere, waiting to be recognized, if only we look hard enough. And once they are found and believed in, instinctively we will shape our future, lining it with the reds and yellows and blues of our pasts.

How my mother lost her leg: I begged. I'd watched
a friend show off her new red jellies all day at school, and I
wanted a pair. In blue. At fifteen years old, I was not immune
to flights of fancy and social pressures. As soon as I got home
from school that day I began hounding my mother.

"Fifteen minutes. Just fifteen minutes!" The store was a
short trip, and I asked her to drive me to where I could buy
the jellies. I'd buy them with my own allowance. I argued
that she'd taken my sister into town the previous day, so
now it was my turn. I wanted them. I was willing to commit
to more chores, better homework performance, anything. I
just wanted them—PVC jellies as colorful and cheap as
they come.

My mother could stand only so much and I trailed her
from the kitchen to the living room to the deck outside and
back into the kitchen. I bargained. She gave in. We walked
together to the car bays where we loaded up and drove the
fifteen minutes to the store. I bought my jellies, blue ones.
She picked up incidentals. Her purchases fit into two plastic bags. Mine fit into one. I clutched it to my chest as we
walked back through the parking lot and climbed into the
Ford Explorer.

We turned out of the lot, merged onto Highway 7, and

nosed back towards Mount Rainier and the farm. One left turn from home, still on the same road, we glided along a straightaway flanked by farmland and forest. I stared at the blue jellies in my lap. I didn't see the other car pull out into our lane, coming at us speeding, in excess of seventy miles an hour. To this day I do not know why a woman named Patience decided to pull out and pass five cars. But we were in the oncoming lane.

My mother veered and took the full impact of the head-on collision on her side of the vehicle. The impact threw our SUV into the sky and rolled us over and over and over. We settled in a shredded mass of metal and plastic in a culvert where paramedics came and jabbed needles in our arms. They took us to the hospital. They helicoptered Mom to Seattle where she woke up the next day without a right leg.

She didn't do physical therapy at the hospital afterward. Angry and bitter, she wheeled home where she sat and waited for her stump to heal. She rocked around on crutches but couldn't use them much when the sores they caused under her arms bloomed and raged. After being fitted for her prosthetic leg, she strapped it on and went back to walking and watering her garden. She had a farm to run.

I remember her, not long after the amputation, sitting in her wheelchair, parked in the old house next to the huge Boston fern she'd nurtured for as long as I'd been alive. She was hunched in her wheelchair, looking out the window, chain-smoking. The day was getting dark outside, and her face blurred in the shadows framed by her black hair. The thick smoke made everything boundaryless. One hand rested in her lap; the other held her cigarette near her lips. In the fading light the only things clear were the silver bracelets stacked up her left arm. They tinkled softly as she moved. I stood in the hallway, watching. She moved, brought the cigarette to her lips and harshly inhaled. The flash from the tip of her cigarette lit her face, briefly. Her eyes glinted.

My mother experienced a transtibial amputation—doctors made the surgical cut just below the knee joint. The head-on impact of the two vehicles, like two tectonic plates colliding, crushed my mother's foot and her lower leg. There was no hope of reconstruction. No mitigating the damage.

There are things I know now that I did not know then. I know that before the doctors performed the amputation, a prosthetist came into the hospital room and measured my mother's smashed leg. The answer to amputation is technology. Even while she was in surgery, the prosthetist started the long process of making her a new limb.

The prosthetist was both a doctor and an engineer. He had to construct, with his own hands, a technological solution to her medical problem. Three to four weeks after the surgery, once the swelling had gone down, a plastic mold was taken of her stump. This mold was used to ensure a good fitting for the socket of the prosthesis. Six to eight weeks after the amputation, she went in for her first fitting of the new limb.

I know my mother must have been exhausted, nervous, worn down at this first meeting—the first in a lifetime of such meetings. She'd visit Dr. B. multiple times in the following months to make sure her first leg fit properly, and then, in the coming years, she'd go in every few weeks for fittings, tinkerings, mechanical malfunctions—and, of course, for new limbs.

Each limb had an average life span of three years. There were definite limitations to the technology. It seemed that each new limb wore out more quickly than its predecessor.

Time and her stomping, striding, walking, strolling around our farm wore them down. When she died, a whole corner of her bedroom was taken up with dusty worn legs leaning in on themselves like a prosthetic house of cards. My father donated them to Dr. B. The National Limb Loss Information Center estimates that one in every two hundred Americans is an amputee. Someone might use her old limbs, especially since individual prosthetics retail at tens of thousands of dollars.

Prosthetic limbs aren't new. Humans have been losing extremities since the dawn of time. In 2000, scientists discovered what they believe to be the world's oldest prosthesis. At a dig outside Cairo, Egypt, archaeologists excavated a 3,000-year-old mummified noblewoman—complete with her prosthetic big toe made of wood and leather.

Since the time of the Egyptian pharaohs, artificial-limb technology hasn't evolved much. Wood remained the material of choice until the sixteenth century, when French surgeons incorporated metal hinges for movement. But the greatest leaps in prosthetic advancement have happened relatively recently as a result of the US war in Iraq. More soldiers are surviving what previously would have been considered fatal wounds than in any previous war; however, they're losing more limbs. For every soldier killed in Iraq, according to the *New England Journal of Medicine*, nine others were wounded and survived.

Now that the US military is accommodating so many more amputees, it is directing funds towards significant research and development in the field of prosthetics. Huge advancements have been made in the field, and amputees, paired with great technology, are pushing through and surmounting immense challenges. Notably, Tom Whittaker, an avid outdoorsman who had his right leg amputated below the knee after a car accident, became the first disabled person to summit Mt. Everest, the tallest peak on the planet. It was his third attempt, and in 1998, he stepped successfully onto the summit.

Whittaker used new technology that stored and released energy to match the force of the regular step of his good leg. Speaking to the New York Times, Whittaker noted, "I've never had anything in a prosthetic device that duplicated the range of movement that an ankle socket can, but this comes the closest. Wearing most prostheses is like walking around with your foot in a bucket."

Whittaker's observations are especially poignant. His earliest prosthesis was indeed a sort of bucket, a large tobacco can he attached and taped to his leg's stump.

I'm intrigued by prosthetics. They're not just plastic legs or arms. Prosthetics can be anything that replaces or enhances a function of the body. If a system or function is not adequate, or fails, a technological solution might be applicable. Such solutions meet the needs of disability, level the playing field. Limbs, eyes, noses, ears. Breasts, lungs, hearts, jaws. Mirrors.

I think our planet is slowly becoming disabled. Due to climate change, in natural processes that digest carbon, regulate temperature, keep the climate on an even keel, everything is off kilter. Undoubtedly, unquestionably, it has been shown with certainty: the way we live our lives is causing these systems to fail.

We talk about the complexities of climate change, my friends and I. The students I work with all ask what is to be done. The newspapers don't record many climate victories. Engaging in any meaningful way with climate change feels tremendously overwhelming. The world we know, the one we trust in and interact with—the basic parameters have changed.

My mother never paid attention to prosthetic technology until she needed it. She was living her life, walking around on two good, strong legs, and then suddenly, the choices of another human being left her disabled. She needed help to be able to live her life. She faced her disability and took advantage of prosthetics to ensure the highest quality of life she could achieve. My mother never wanted the very best prosthesis; she just wanted her body to function again.

Geoengineers could be called prosthetists for the planet. They're scientists exploring technological solutions for Earth's planetwide system failures caused by climate change. Their research, geoengineering, has long been described as akin to planetary Tylenol—not a solution to climate change unto itself, but perhaps something that can take the pain away. A technological pill for the pains of human behavior. It's the breeding of clouds, inserting of high levels of aerosols into the atmosphere, treating the oceans with iron, sequestering all that carbon into ground tanks. There are so many geoengineering schemes. But it's considered highly problematic. I can suture my head every time I bash into a wall, but the suture won't stop me from bashing my head in the first place. The damage caused by continuing and unchecked greenhouse gas emissions is barreling along at such a rate that environmental measures such as energy-efficient light bulbs really don't appear adequate.

Geoengineers have been accused of attempting to manipulate nature. They often confess wholeheartedly to this charge. They argue that getting human beings to change their behavior is going to require an act of God. And while many in America are indeed waiting for divine intervention, what geoengineers have access to on Earth is science and technology, and they want to use it. The planet is in serious trouble, they remind us, and it appears unlikely that humans will take the initiative to control their greenhouse gas emissions. Therefore, geoengineers' goal is to transform nature on a planetary scale, aiming for positive environmental change as the target rather than the side effect.

In the aftermath of the 2009 United Nations Climate Change Conference in Copenhagen, when many individuals, organizations, and nations realized that planetary agreement on the facts of climate change was not yet universal, geoengineers began ramping up research efforts. It is clear now that no single solution to climate change is within reach; rather, solutions will be varied and complex and include many different treatments.

"Climate policy," writes David Keith, a leading geoengineer, "is often framed as a choice among various energy technologies and policy instruments. Beyond this choice of tools, however, lie hard choices about the objectives of planetary management. . . . A strategy of active management might freely employ a mixture of responses, including the reduction of CO_2 emissions, geoengineering, and strategic adaptation to changing climate."

It is possible that human society's existence on this planet is at the cusp of a paradigm shift where our greatest disability—the damage we have done to this planet—could be the catalyst for developing our greatest ability: the human capacity to imagine, learn, and solve problems. This "disability" might be the dark and deadly catalyst human society needs to begin changing how we imagine living our lives in this world. If nothing went wrong and climatic changes were not happening, undoubtedly we'd continue business as usual.

But today, it simply is not possible to look around at the natural world we inhabit and avoid seeing the immense environmental destruction and degradation we have caused at local, national, and planetary levels. If human society chooses to do nothing, the consequences are unimaginably grim. Climate change is playing for keeps. Here is a great opportunity to re-imagine. We can do better, and we should.

Human societies—you, me, us—are at a disadvantage now, and we have the option to sit back and watch and do nothing. Or, we can allow ourselves to measure the extent of our broken pieces and imagine better ways to live, some of which might include reaching for, and using, the newest and most current prosthetic technology in order to get us to the real end, the real solution: a world with dramatically reduced greenhouse gas emissions.

I never fully appreciated my mother's feet until she had lost one of them. She had had what I thought were perfect feet. They were small, size five or six, depending on the shoe. She had a spring ritual in which she would buy herself a new pair of sandals—it was common to hear her ask, "Where are my flops? Anyone seen my flops?"—and throw away the previous year's worn-out pair.

She used to laugh at my feet. I received from her my orderly, picket-fence minor toes, but my big toe was inherited from Dad. it looms monstrously two to three times larger, casting shadows over my entire foot. I often caught Mom's big toe casting sympathetic glances at my ogre-toe.

My big toe catches in the hem of whatever pants I am wearing—it happens so quickly—and I am walking and then I am on the ground. Several years ago I was walking through the living room at home and my toe caught on my mother's oxygen cord. I promptly careened over, smacking into the rocking chair, pulling the nasal cannula out of her nose, and nearly causing her to topple. There was a pause as we gathered ourselves. No blood. We laughed then.

Once, years before she was leashed to the oxygen machine, we were both out in the front garden, tending the corn and sunflowers. I remember looking over, seeing her there in the yellow flowers, and then looking away. Next time I looked, she'd disappeared. Balance is difficult to achieve on a prosthetic leg, especially on uneven surfaces, like garden dirt. She'd tripped over her own foot—the good foot, the one that still had the perfectly aligned toes—and had gone down in the row of plants. There were tears in her eyes when I rushed over. Frustration tears, not pain tears.

Mom used to fall a lot, especially in the beginning when she was getting used to walking again. Doctors told us we were supposed to just let her fall. Seeing her, on her butt in a sunflower row, fighting tears that were an amalgamation of frustration, fear, anger, and pain—there were a lot of things I could have said, words and kindnesses to make the situation bearable. I should have joked, offered support. But I didn't. I was immobilized by all that I could say, so I said nothing. I didn't hold a hand out to help her; I didn't squat down and join her in the dirt. I just stood there, looking down, blocking her sun, awash in guilt. She got up, like she always did, and she wiped her face, and we went back to gardening. She didn't say anything. Neither did I.

She couldn't wear flops after the amputation. Her prosthetic leg didn't have a gap between the big toe and the little toes. The first spring after her amputation I started zeroing in on the orphaned pair of flops that rested by the doorway. I stared at them every time I went in and out of the house. They were dingy, shoved back by the mud-boots, but visible even though there were shoes piled helter-skelter. At that time, Mom was still trying to figure out which shoes she could wear, which she could get on and off her plastic foot. She was limited to a one-inch heel or less, and trying on shoes in the store was a trial she often didn't have the energy for.

Once, in J.C. Penney, she got a boot stuck on her prosthetic. It was a black, ankle-high boot that zipped up the side. Mom put the boot on, zipped it, and stood. When she sat back down and tried to take it off, the boot wouldn't budge. The clerk yanked and pulled, but it wouldn't move. Finally the clerk wrenched it off with the biggest shoehorn I'd ever seen. Mom sat rigid with embarrassment. I hid.

That evening, back from J.C. Penney, I stole her flops as the fog rolled in. Dad had lit the trash barrel down by the barn. Smoke drifted up and I smelled it in the house. Mom stood in the kitchen, leaning on the counter and putting together dinner. The smell of the smoke mingled with the aroma of early dinner. She didn't see me as I slipped from my bedroom, walked behind the couch and prepared to go outside to clean the horse barn. I usually told her I was going out and checked to see if she needed anything; I didn't that evening. By the door, putting on my rubber boots, the shoe pile shifted and I made eye contact with her flops. They were pathetic. I was sick of seeing them. I grabbed them, slipped them under my jacket, and left the house, walking down to the barns.

Dad was multitasking, moving between the barn and the barrel as he fed the cows and tended the fire. The Angus cows pushed and shoved, their deep lows reverberating off the barn and disappearing into the foggy pastures. Steam wavered off the cows. The light was dim, blurry. Smoke and fog mingled. I could smell the heaviness of damp trees. I stood near the burn-barrel.

I waited until Dad had his back turned, and I chucked the flops into the flames.

When David Keith came to Missoula for a geoengineering conference, we spoke, briefly, about his plans to manipulate the planet. He's a busy man; even more now, I imagine. The United Nations Convention on Biological Diversity met in 2010 in Nagoya, Japan, and the meeting closed with the attendees in agreement on a moratorium on all geoengineering projects and experiments. Silvia Ribeiro, director of the international human rights and sustainable development organization ETC Group, which is famously outspoken against geoengineering projects, reported: "Any private or public experimentation or adventurism intended to manipulate the planetary thermostat will be in violation of this carefully crafted UN consensus."

When the UN banned all geoengineering projects, David Keith went through the roof. He believes that continuing research and development will buy our planet more time to mitigate climate change. His stance is clear: "Climatic geoengineering aims to mitigate the effect of fossil fuel combustion on the climate without abating fossil fuel use; for example, by placing shields in space to reduce the sunlight incident on earth."

This is the time and place for bold thinking. I think audacity should be applauded, encouraged. When stepping into the climate change debate, we need bold thinkers, inspired men and women who create and nurture small ideas, big ideas, and everything in between. There is courage found in imagining and re-imagining different paths forward.

As a scientist working on climate change, I don't think geoengineering the planet is a good choice. Solar radiation management, cloud seeding, artificially inserting aerosols into the atmosphere, ocean iron fertilization, and so on: this list of proposed geoengineering projects is long and imaginative. But at its core, such projects remain planetary Band-Aids that fail to address the core problems that feed into the current rate of accelerated climate change. That being said, I do think that a conversation about geoengineering should continue. The ban on geoengineering is stiflingly unimaginative. If we allowed research into this area—if we welcomed the possibility of saying, "what if?"—couldn't this provide a space, a catalyst, for nurturing and developing great ideas, for building a universal environment that encourages individuals and groups across the planet to re-imagine boldly what the future may hold?

I've never been much of a gambler. I think that as of this moment, climate change needs to be addressed from many different perspectives. We're too early in the process to completely dismiss one whole field of research offhandedly. It takes a great deal of audacity to stand up and say, "How about this?" If we set a precedent that indicates we're not willing to study, to ponder, the most cutting-edge of ideas, the boldest of thinking—I shudder to wonder what avenues may be arbitrarily closed off.

Climatic changes are not going away; rather, the environment human beings prefer to live in is. The planet's most vital systems are not failing so much as they are shifting rapidly away from what humanity has counted on for centuries. Services we depend on for the air we breathe, the food we eat, the wildness that assuages our souls. If we halt all geoengineering research and development, what happens if the very worst comes to be? What happens if the planet finds herself driving down a lonely stretch of highway in mid-afternoon with her child clutching a pair of jelly shoes to her chest?

Geoengineering research allows us to possess more arrows in our quiver in preparation for the worst-case scenario when we actually have to shoot at something. Additionally, it is entirely possible that, in the process of researching one of these technologies, other undreamed-of arrows could be realized. In essence, we have the potential to research advanced planetary prosthetics, but we are limiting ourselves basically to a wooden peg and an eyepatch. Stopping now, with the scantiness of the research that has been conducted, is akin to a lobotomy. By abandoning, dismissing, or condemning the possibilities of geoengineering, we limit the range of possibilities. Recall David Keith's comment: "shields in space."

"Shields in space" translates into solar radiation management: reduce the amount of solar radiation coming to Earth by reflecting it back out into space. Less greenhouse gas effect, slower climatic changes. But continued, or increasing, greenhouse gas emissions.

I admit it, though, shields in space divert my attention immediately, and I envision a large umbrella parked over Earth with solar raindrops dripping off the edges and falling into deep black space. I'm actually not far off, excepting the scale of the project. Imagine a single umbrella over a thousand miles wide (about one-eighth the width of Earth) composed of sixteen trillion small, reflective discs. For as long as necessary, Earth would be shaded from the sun, buying our world more time to wrestle with our greenhouse gas-emitting lifestyle.

The estimated cost of the project, which NASA researched, surpassed five trillion dollars—a number I can't conceptualize. Along with the staggering cost, there was also the question of how to get an umbrella an eighth the size of Earth off the planet's surface and into space. Suggestions to date include building the shade incrementally and flinging each piece into space with a railgun, or, more likely, constructing the entire structure in a factory located on the moon. The scale of the project fascinates me.

Estimates project a 2 percent decrease in solar radiation after the shield is in place—enough to significantly slow increases in global warming. In effect, this global prosthesis could stop dangerous climate change. Temperatures globally would drop down to pre–Industrial Revolution times. It would buy us enough time, as David Keith stresses, to get our global act together and cut back emissions.

I am not necessarily comfortable with this as *the* backup

plan. If we, as a species and a planet, find ourselves in such dire circumstances that we must employ a geoengineering solution, then it means we cannot first address what caused the problem: greenhouse gas emissions. What use is it, then? I appreciate David Keith's point that a temporary prosthesis would be a way out of a pinch, but it shouldn't be the solution unto itself. My mother is not her prosthesis. Rather, it is her will that directs her prosthesis to help her as she moves about the business of life.

My father knows he has cancer. He watched my mother die. He saw what that was like. And he wants options: all of them. The alternatives, the hare-brained schemes, the traditional avenues. The trusted landscape of his world is sliding away, and he sits up late in doctors' offices, listening, talking, imagining. What experimental drugs are out there? What do they do, and, importantly, how much do they harm? What are the ways forward? To him, the word "terminal" is not definitive; rather, it is merely descriptive. My dad knows he still can define his own future; he still can imagine his preferable paths forward.

Oddly, when I think over geoengineering projects, my mind wanders and I imagine my mother waking up in Harborview Medical Center in Seattle. She's groggy; she's cold; she can feel tubes and ports and needles; she hurts all over. And she knows, she knows before she looks, that something is missing.

When she does look, down towards the end of the bed, she sees her left foot, the silhouette of toes poking through the white blanket. The problem, her brain slowly registers, is that there is only one leg down there. One leg. Where the other leg is supposed to be there is only smooth blanket. Nothing.

The knowledge came first, I imagine: before the pain. And when the pain came, it knocked her back into the hospital bed and embedded her into the pillow. She had lost her leg. Months after the car wreck I concentrated on the word "lost." I was a teenager and didn't know anything about medical procedures. I understood my world in terms of binaries. I conceptualized "lost" as the opposite of "found." I wondered where my mother's leg went. If her leg were lost, could it be found?

How does one wake up without a leg? There are choices to be made, new landscapes to understand. Before driving home from the store, the topography of our lives was known and expected. Now, a one-legged woman struggled to negotiate steep mountain slopes. My mother had to relearn her life. Simple things, like standing, getting out of bed, rising up on tiptoes, showering, running across a lawn to snatch a child away from a backing-up truck: these things were no longer automatic.

In the clear light of day, I can intellectualize that I am not to blame for the loss of my mother's leg. But then, at night, I doubt. And what I know is that because of my shortsightedness, my desire for something insignificant, I put into motion events that would have hellish repercussions for my family and the remaining years of my mother's life. There is no isolated act. I am not to blame for the car wreck, but I am part of something larger, a collective fabric that ties us together. We all are. The decisions, the dreams, the imaginings of each individual person—gradually, in known and unknown ways—reach out and touch each and all of us in turn. The very audacity of our dreams is indicative of the collective condition of our human spirit. I am reminded of this while standing in the shower, staring down at the water running along my legs and between my toes.

Patience was never held accountable for her actions. Police at the wreck scene failed to administer a drug or alcohol test. Why this failure happened, I never fully understood. She paid a minimal fine and walked away.

I find myself immobilized in the wake of her choices, of mine. In the months after the car wreck, while my mother healed, I raged at her. I didn't participate in her limb building, the doctor visits, the construction of ramps and the helping with chores. By doing nothing, I believed I was somehow avoiding responsibility. I felt that if I did something—if I helped or moved or cried or brought in the last bag of groceries—I would be admitting and assuming the mantle of a wrongdoer.

What would my family have thought of me then—the one who so desperately needed a pretty pair of shoes? Would my dad have been able to look me in the eye? I remember, as a teenager, sorting it out in my mind: I was responsible. Therefore, I could not take responsibility. Almost like, today, it is widely known which countries emit the greatest amounts of greenhouse gas yet will not take on the burden of admitting it and moving forward. I wonder, looking back at myself, whether I did not take responsibility because I could not imagine what that looked like; I could not imagine what I could do to make it right. Today, looking back at my teenage self, I see countless parallels between the way I responded and how people are responding to climate change. We seem so busy not being responsible.

My mother and I rarely spoke about the car wreck, and if we did, it was always about what happened afterwards. Slowly, over the years, the language in my family changed. We no longer say "car wreck." We say "car accident." When we speak of family events, family timelines, often it is time-controlled by noting before or after "the accident."

But what happened that late afternoon on Highway 7 was not an accident, not in the pure sense. It was the product of individual choices. At fifteen years of age I could not fathom the decision of another driver to pull her car out of her lane intentionally to pass five other cars. What she was thinking, I have no idea. No one intended for that to happen. But it did.

I think my mother bore the trauma of her loss in part because, as she sat in her wheelchair, chain-smoking, she imagined her options. That in a lab somewhere, an engineer was assembling her freedom. And that the past choices of someone else did not have to define her future. She was limited by her imagination, by her own dreams, by her own courage.

She went north and taught herself to walk primarily because she imagined herself walking again. She took a brutal blow and moved forward.

I REMEMBER CLEARLY THE DAY I WAS TOLD MY FATHER
was dying. It was my brother who called, who bore the
weight of sharing this dire news. I was en route to the
University of Montana, walking, headed to a meeting with
my thesis advisor. I was on the home stretch of graduate
school, putting together the final timeline for my thesis,
ready to accept my master of science degree. I'd applied for
and earned a US Fulbright grant to travel to Turkey for the
year after graduate school. I was planning on enrolling in
a doctoral program at the University of Oregon to further
my research on glaciers and climate change. It felt, walking
towards that meeting, like a triumph of sorts. My mother
had died when this all began, and, a year and a half later, I
was still there and close to successful completion. The future was bright.

My dad had been sick, diagnosed with cancer, but there
had been hope, treatment. He'd been responding well to an
experimental cancer drug, Sutent.

I walked and walked with my head focused on the future
and my phone rang and then it was raining.

Or rather, I expected it to rain, to flood, to wash down all
possibility of this world being real. It rained in my mind. The
kidney cancer in my father had spread throughout his body

to his lungs and there wasn't anything the doctors could do and the Sutent had stopped working and and and.

Time in the form of uncountable hours passed, and I was on the phone, calling a friend whom I hadn't seen in months, and he was there on the side of the road, and, picking me up, he drove me home. And then, later that day, he and I were driving in my car, travelling late into the night, driving out of Montana, traversing the narrow strip of Idaho, and then entering into the moonscape of Eastern Washington. I'd made this journey before. Last time, in the heat of grief, I was on a plane in the late morning, flying to my mother's bedside. Now, car barreling across black asphalt late at night, I was heading to my father's side, the fragile landscape of my world crumbling, again. How do we survive this?

The last barrier remaining between the farm and me was the looming Cascades, a mountain range best crossed at night, when the mind has time to commune with the greatest of earthly forces. The Cascades: they house gods, I'm sure. Their name and number remain a mystery to me, but I know, looking at the mountains, that there is more there in their tall, jagged peaks than what meets the eye.

I prayed bitterly to those mysterious gods as the mountains emerged in the periphery of my headlights.

I cracked a window in the car, just briefly, and the car flooded with the smell of salty wind transported from the Pacific.

I remember when a friend of my parents went to Japan during my childhood. She returned with presents, and gifted me with a small postcard that depicted Fujin, the Japanese god of wind. The picture terrified me because it had a large man with an angry expression reaching down from the skies and swiping at the lands below. The man appeared to be funneling out of a large volcano. My child-self burst into tears.

The woman who had given me the card knelt down. She explained to young me that Fujin was only angry because he missed his family. All wind gods, she said, lived on top of Mount Rainier, the biggest peak in the Cascades. Surely not the woman's intent, but I cried harder, sad for Fujin, that he had lost his family.

From the farm, just over the tree line, Mount Rainier heaves upward into the sky. I grew up knowing the topography of that mountain, the placement and change of her glaciers, the names and sizes of the various unique cloud formations that would sag into her high peaks and dissolve down the eastern side. I used to look at the mountain and wonder why Japanese gods lived there.

Driving west, gaining elevation as the road wound up the mountain's skirts, I prayed and prayed to any god living in the mountains to change what I was experiencing. I begged them to change the throws of fortune, to alter the outcome and not lead my family down this dark path once again, so soon after the passing of my mother.

The gods, the mountains, they all stood still. Small lights blinked red outlining the rotor blades of the numerous turbines blowing in the Cascades' skirts, but no word from the gods.

The year before my mother died, my father was diagnosed with cancer.

"Dad," my brother told me, "sat in his rocking chair for days after the diagnosis. I've never seen him like that. It shook him."

Dad was not a lounger. His rocking chair sat in front of the fireplace next to my mom's knitting station on the couch. His chair was occupied in the evenings when he read the newspaper or a magazine, or watched a movie with my mom, or perhaps, on the rainiest of days, he'd perch briefly in his chair and assess a football game before heading back outside. My parents built the tiniest of farmhouses because their entire lives were spent outside.

For him to park in his chair meant that something was very wrong.

But then, three days later, Dad got out of his chair. There were cows to feed and a family to look after. Doctors were optimistic. There were new drugs open for trials. Dad responded well to them.

Even today I cannot help but marvel at how many good, though imperfect, solutions to the many problems of climatic changes already exist. Some of these solutions provide visible steps, positive signs, forward motion—and though they are not the perfect answer, they are progress. Like Sutent.

After Mom died, my brother and sister and I shifted gears rapidly: the focus went from my mother to my father. How do we keep this man going? His body stopped responding to the drugs doctors prescribed.

Dad's oncologist suggested we try Sutent, a new drug from Pfizer that, especially in comparison to all the other drugs, had negligible side effects. It was designed specifically to address kidney cancers. Dad took his small pill every day, and it was a miracle: his cancer slowed. Then stopped. And for months, for more months, Dad lived his life. He wasn't cured of cancer, but he was living successfully with it. A temporary answer.

He stretched out and made a life that was beautiful to see. He'd lost his life partner, my mom, and he grieved. But he also made new friends, and he volunteered, and he travelled, and he visited me in Missoula, and he got upset at Republicans. He discussed the price of hay with our neighbors, he considered which cows he'd buy from the Angus auction in Eastern Washington, and most mornings, he stormed out of the house wearing his bathrobe while waving a mop at the woodpeckers drumming at the cedar siding.

I think back to the late-night, dark, joyless car ride that whipped me from Missoula, Montana, to our farm in Washington, and instead of music, or conversation, or tears, what I remember most are the huge, ghostly windmills that ripple steadily above Interstate 90 near the Columbia Gorge. Undulating answers.

I should have been focusing all my energy on my father during that drive; instead, I was strangely comforted and awed by those sentinels. They flooded my mind, the steadiness of their movements evoking within me a calming curiosity.

Driving west from the Columbia River, past the Ginkgo Petrified Forest State Park, the land rises up on either side of the highway. Rooted on the tops of those mirrored hills are giant, 280-foot windmills, their rotors turning metronomically in the breeze. They weren't there ten years ago. Today, the windmills dot the eastern slopes of the Cascades like a chessboard. If you squint your eyes, they look like a multitudinous army of white, metallic giants with blurry halos, marching, marching. They remind me of the Imperial Walker machines operated by the Galactic Empire in Star Wars. Hoth, recreated in Washington State.

Washington, however, is a bit more advanced than Hoth. It has an energy policy that includes renewable-energy standards requiring all major utilities to incorporate 15 percent renewable energy in their portfolios by 2020. Owing in part to this, and in part to extensive federal tax breaks, the Pacific Northwest has seen a surge in wind energy, doubling the amount of electricity generated by these farms in just three years. The energy is used locally and more widely: roughly 47 percent travels south along the West Coast to California utilities.

To me, the windmills give off an air of hope. They are tall, narrow, rather minimalistic in shape. They appear healthy, fit. Each windmill's rotor blades span four hundred feet and weigh approximately fifteen thousand pounds. They only need wind speeds of six or seven miles per hour to turn, and each generates ninety megawatts of electricity.

It's a small leap to think that, if the windmill were alive, it would like living communally with its wind farm neighbors; I bet, if the windmill could, it would shop locally, buying wine from the burgeoning vineyards, and likely, it would listen to NPR while shuttling its windmill children around in the family car.

The windmills, of course, need wind to energize themselves. Wind is palindromic action, an expression of the vibrant energy surrounding, lining, permeating every atom of our existence. Wind is a blip in the matrix, a bright advertisement on a dark night. It lends itself to possibilities and far-flung places, to the building up and tearing down of mountains, to the buoyancy of hope in midafternoon. It is, undoubtedly, the voice of the sun.

Wind is born from the benevolence of the sun. She shines down upon our planet day after day, but is rather unequal in handing out her heat. The ground surface, the skin of Earth, heats unevenly based on widely disparate parameters, including elevation, latitude, terrain, dirt, water, and ice. Add in the complications from the rotation of the earth, the heating of the atmosphere, and the pitted nature of the planet: these all exacerbate, encourage, and help create energy through wind.

It takes a village to raise wind.

Like water, like love, there is only a finite amount of energy in this world. Wind is a constant expression of this energy.

Wind farmers are energy dreamers of a sort. Wind is not the easiest, most tangible thing to see. Wind farmers have to imagine the wind, imagine their crop. They place their windmills in sites they think are the most likely to tap into roving energy—like, say, a beautiful woman positioning herself in the most attractive light to better lure a lover. As the wind flows by and through, wrapping the planet again and again, air is funneled through wind turbines and used to generate electricity, which then feeds cities and homes and cars and laughter and love and disperses back out into the universe.

It is not possible to drill for the wind, to blast deep into tar sands and extract playful energy. It is not possible to pump your car full of wind or run a nuclear reactor on wind alone. It is silly to buy wind, or try to trade it on the world market. Harnessed wind is not wind at all; rather, it is chains on the ground surrounded by vacant laughter. Wind cannot be abused, or depleted, or made to spew the most ugly of toxins. Wind wishes you a good night; wind is boundless and capricious. Wind will not come when called; it is only colored by our imaginations; wind is profound. Sourced from the sun, wind spins our planet like a pinwheel.

It is not possible to miss the wind when it is gone because it is never really gone.

In 1973, Ernest Friedrich Schumacher, author of the delightful book *Small is Beautiful: A Study of Economics As If People Mattered,* wrote, "Perhaps we cannot raise the winds. But each of us can put up the sail, so that when the wind comes we can catch it."

We've been putting up the sail for a long, long time. In the seventh century, the first practical windmills were installed in Afghanistan. Blocky, with reed matting and cloth sails, these structures operated long, central driveshafts to pull water from the earth and grind down grains and sugars.

Centuries later, the task of the small windmill on my family's farm is the same: to pull water.

The windmills atop the Columbia Gorge are similar in structure but different in task. Imagine a big fan, but flip it around. Instead of a propeller, it works as an impeller. A fan uses electricity to make wind. In a windmill, the spinning blades rotate a shaft connected to a generator. The energy output is electricity.

Now imagine bigger, grander. Dream. Envision floating high over the earth, but still in the atmosphere, where fierce winds rage above the planet's surface. Imagine huge fans, or sails, floating on those currents, spinning and spinning at such speeds they look like twinkling stars just within arm's reach. Now imagine those windmills of the sky anchored back to earth, where they feed into ground networks. Imagine harnessing the sky.

I am attracted, perhaps, to the windmills situated around the Columbia Gorge because they appear to be upgraded versions of the windmills from my childhood. My dad, long before he started working half of every year in Alaska, long before Mom was ever sick, long before I acquired any type of indelible memory, dreamed of being a windmill maker.

The windmill on our farm is one of its most prominent features, the forty-foot landmark we give friends who are trying to find us. Drive one mile up the paved road. Look to the left for a red mailbox and a huge, red windmill. Can't miss it.

When my parents bought our farm, they decided to build a windmill to pump water. My father was a pipefitter, a craftsman who worked with iron. The farm's windmill would be the first of many he'd build. It took him several weeks to construct the tower, hauling iron bits from the Port of Tacoma to build it higher, higher. He ordered the motor from Ohio, eagerly waiting as it was shipped in bits and parcels. Each mechanism arrived separately.

They laid it out on the ground: each level, the rotors,

the tailfin. The motor. Mom held the rods while my dad welded.

One evening in the summer, with the hazy light of the setting sun casting golden hues on the surrounding pastures, my parents raised the windmill themselves.

I imagine the triumph of that moment. I know what it looked like—even though I was not there. I know that my dad had long, blond hair that billowed around his strong shoulders. My mother had brown-black hair, pin-straight, that reached the top of her Levi blue jeans. I know that they both wore muddy boots up to their knees, that my dad probably wasn't wearing a shirt, and that the bracelets on my mother's arm jangled as she worked. I know they periodically stopped, had a coffee, shared a smoke.

I imagine them standing there, next to the pond, with rigging and tractors and all sorts of confusion. Both holding coffee cups, discussing the next step.

"Scaffolding is finished," my dad says.

"Took all day," Mom responds. She laughs.

"Thank god the Hunters lent us the crane. We wouldn't have been able to rent one."

Mom nods. They stand together. They sip their coffee. The evening light is dusty.

"Well, Cath," Dad says, "let's get to it."

They part ways, my mom to the truck, where she's operating the crane, and Dad to the head of the windmill, which is lying on its side. It's time to put it up top.

I know my dad must have looked at the rotors, at the fins. And he must have taken a second to admire Mom's handiwork: she'd painted the name of the farm onto the tail of the windmill. In multiple colors. Reds. Blues. Yellows. Greens. She has paint on her fingers.

"Nice job!" he yells to her. But she can't hear him— she's got the truck roaring and the crane ready. It's okay; they have time.

For several years after that, Dad travelled the county building windmills. But he was too early with the technology and there wasn't a market and he couldn't feed his family, so he turned to other things.

It's tall, our windmill, about forty feet, standing guard over the farm's home acre: the pond, the barns, the house. It is subtle that this windmill is the fulcrum of activity on our farm. Our family's eyes fell on this structure hundreds of times each day, and we all noted the speed of the rotors and which way the boom was facing, yet rarely did we mention such things. The windmill was part of the woven fabric that held us in, held us together on the farm. The air we breathed spun the rotors, connected us.

The windmill's tail is painted like a rainbow. My mom drew on the rainbow, the blues and reds and yellows, after they'd decided to call their home "The Rainbow Farm." The name was significant because my parents both found hope in rainbows. They found hope in their first piece of land, their farm, and the life they imagined before them.

Over the last thirty years the paint has faded, but if you look closely, you can still see the rainbow outline. The windmill itself is a bit of a rainbow, with different-colored paint jobs evidenced at varying heights on the tower. Summers on the farm, for as long as I can remember, always involved massive painting efforts. Dad had a fondness for John Deere Green, which we slathered all over our corrals and gates and each other. The windmill, however, was never allowed a green hue; instead it was always cloaked in a heavy coat of Beacon Red. But one summer, I had to scramble up there with a wire brush and grind away all the lichen and old paint and rust before we could actually paint it. Then another summer we laid down a primer coat, and another painted the entire thing. I was never a good windmill painter, however—too easily distracted by apples to count and ants to stalk. Years ago we considered painting the windmill one solid color, but that idea never seemed to cotton on at our farm. Hence, even today, the lower stories are primer black, and the upper stories are red. A reminder that nothing is ever finished: that everything is a work in progress.

As a child, I used to scamper around the windmill. I've never, even to this day, climbed to the very top of it, but I used to inhabit its lower stories. A very small, very questionable platform perches at the top of the tower where the motor is attached. The courage necessary to balance on a thirty-year-old wooden platform buffeted annually by Pacific Northwest precipitation was beyond me. The windmill is built like a scaffold, with several steel beams squared together and connected to the layer above it, which is slightly smaller, so the entire structure scopes into a narrow point up top. Even to a person standing on safe ground, peering upward through the telescoping innards of the windmill, that tiny platform up top positively reeks of rotten wood.

I never had my own treehouse as a child; my brother had the market cornered in that department and owned prime real estate in all the best trees. But still I strove for great heights. Initially, I used to haul out my parents' orchard ladder into the middle of the lawn, set it up, then climb to the top and sit for hours, peering around at the world below. There are times today when I chuckle to myself—what must my parents have thought of their odd youngest child? Some days, I'd even bring my cat along for company on my high perch. But the problem with orchard ladders is that you can't really move around much on them. I had the height, just no platform. Sitting on the top of a ladder in the middle of a yard makes one feel silly after a while. So I compromised.

I found a large plank in the horse barn, hauled it to the windmill, and pitched it over the lowest beam I could reach, about five feet high. I hauled myself up, set up camp, sprawled on my board, declared myself home. For years I haunted the windmill, dragging books and troll dolls and music and my cat to sit on the board, bask in the sun, and enjoy my high lookout. When Dad would visit, his eyes would be at board-level, like a floating, talking head peering into my world.

He would check in, assess the state of the book I was reading or how bleached the hair of my troll dolls was. Most often, he'd reach out a hand, a huge, disjointed limb floating from the abyss, and swipe my cat off the board. Her mews as he lifted her down from my perch would set me begging for her return. Dad would free her on solid ground and we'd watch her evaporate into the reeds surrounding the pond. Her fondness for high, solitary places rarely matched mine. Even today I can still remember what my dad looked like from up there, just a head bobbing along.

I can't remember when I stopped retreating to the windmill. It must have been when I reached my mid-teens, but some days it feels like just yesterday that I was five feet in the air, safe in my windmill fort. So vividly in my mind remain the sounds of the windmill: I can summon instantly the sounds of the rusty, crackling driveshaft, the tinny *pop* of wind catching the metal tail fan.

The windmill's main purpose was to pump water up from our well. Running the entire inside length, from the top blossom of the rotors down to the ground, was a small pipe with a wooden dowel attached to it. When the wind picked up, softly cruising the low grounds, the rotors would spin and the dowel would clack and stress metal on metal, shrieking, up and down, up and down, down, down. I'd be careful of finger traps; Dad had long ago pointed out the quickest places I could forfeit a finger. I judiciously kept my digits tucked when the windmill pumped—at times, I even tended my cat's tail to keep her safe.

If the wind was too high, too stormy, there used to be

a latch down on the right side of the windmill's base that could be attached to a steel loop. This would harness down the rotors so they wouldn't spin, locking them in place for safety. In later years, when my siblings and I moved on to college, and my parents moved out of the prefabricated home we grew up in and built their tiny cedar house by hand, they didn't need the windmill to pump water from the well continuously. They removed the pipe that ran from the spigot to the pond that managed all the overflow water. The rotors spin freely now, moving and twisting at will.

In effect, the windmill was neutered.

The farm felt quieter then, not so vocal about the power of the winds. I don't think we knew what we were losing until long afterwards, when the quiet set in.

"Prophesy to the wind, to the wind only for only/ The wind will listen," writes T. S. Eliot in 1930's' "Ash Wednesday." How clearly did Eliot prophesy the future: in 2004, Princeton University professors Stephen Pacala and Robert Socolow proposed, in *Science* magazine, the creation of a five-decade plan exploiting the best technology available today to combat carbon and reduce global emissions. Their plan, set from 2010 through 2060, called for, among other things, the construction of one million large wind turbines (fifty times the current production capacity). In addition, another half a million wind turbines would be needed to support the electrical needs of hybrid cars.

This plan was considered a pipe dream in 2004. It was well known then that wind farms were prohibitively expensive to set up, often so much so that investors couldn't source even marginal start-up capital. But wind energy, once up and running, requires few costs to maintain, cementing its reputation as one of the cheapest green energies available (after setup). Few investors in 2004 piloted programs in the United States, but in other countries, most notably in Europe, wind energy soared.

Today, Denmark accounts for about 20 percent of its countrywide energy needs by wind power, and aims to double that percentage in coming years. Half the wind turbines available on the world market are created by Danish businesses. Investors in the United States, while initially slow to the table, are beginning to take notice, but not without reservations.

Wind farms do have downsides. Free, benign energy does not exist in our world. Any source of energy has varying costs, some of which might not be immediately realized. Benefits and costs must then be weighed. For instance, what level of environmental damage is acceptable to continue extracting oil from the earth? In Canada, the nation has decided that processing a boreal landmass six times the size of New York City to extract the sticky tar sands underneath is acceptable environmental damage.

It appears that, in the United States, there is conflict between wind farmers and bird advocates. Birders argue that wind-farm turbines cause large numbers of bird deaths. Wind farmers say the numbers are marginal. This then begs a cost benefit analysis. A US Fish and Wildlife report published in 2009 estimated the annual number of bird deaths from wind turbines to be approximately 440,000. What is or is not considered "marginal" appears to be open to interpretation. But, for perspective, house cats are charged with the deaths of anywhere between 1.4 and 3.7 billion birds annually—just in the United States. God knows I've seen my own cat drag in enough birds to validate such a huge charge. More perspective: scientists estimate that there are over 200 to 400 billion individual, flapping birds busy minding their own bird business.

Then Interior Secretary Ken Salazar acknowledged the bird versus turbine issue: "We have a responsibility to ensure that solar, wind and geothermal projects are built in the right way and in the right places so they protect our natural and cultural resources and balance the needs of our wildlife."

The American Bird Conservancy's vice president, Mike Parr, responded, "Just a few small changes need to be made to make wind bird-smart, but without these, wind power simply can't be considered a green technology."

According to the American Bird Conservancy, wind farms should adopt bird-smart policies that include "careful siting considerations, operation and construction mitigation, bird monitoring, and compensation to reduce and redress unavoidable bird mortality and habitat loss. . . . Wind farms should also comply with relevant state and federal wildlife protection laws such as the Endangered Species Act, Migratory Bird Treaty Act, Bald and Golden Eagle Protection Act, and National Environmental Policy Act."

Currently, bird-smart policies are voluntary and different for each individual wind farm. Comprehensive guidelines and legislation are needed as wind energy grows across the country. The US Department of Energy is trying to fast-track clean energy, with goals for 80 percent of energy sourcing to be clean by 2035. Wind energy would make up a substantial amount of clean sourcing. However, the American Bird Conservancy warns that such a goal could cause up to a million bird deaths annually.

In 2009, only 2 percent of the electricity generated in the United States came from wind farms. With new wind farms installed since 2012, wind production is projected to constitute 20 percent of America's energy by 2030. Farmers and landowners are becoming more and more interested in this lucrative market because they don't have to change the traditional uses of their land. Farmers can graze their cattle and raise their crops right up to the bases of the windmills, exercising multiple uses for their property.

Wind energy is revitalizing tiny communities across the nation. In Bickleton, Washington, south of the Columbia Gorge, a new ambulance, a new fire hall, a 10.5 million school, and a new town park were all funded by property taxes flowing in from local wind farms. Residents of the area, mostly dry-land wheat farmers, are reaping the benefits of renewable energy by leasing land for rights-of-way, easements, tower pads, and other industry associated with the wind farms. With this influx of energy revenues, suddenly, as Bickleton School Superintendent Rick Palmer said, "We'll start to look more like a community and less like a wide spot in the road."

I've only driven through Bickleton once, and I don't remember it much at all. But I drive through the Columbia Gorge regularly, and there are times, as I slowly start to crest a hill and see the wind turbines spinning atop the ridges, that I wish to stop my Subaru, pull over to the shoulder, get out, and hike up the hill to the base of one of the great giants. I wish that I could stand in its soft breeze, to windmill my own body, fingers splayed, arms outstretched. There must be some moment when my feet are rooted into the ground and my face is tilted upward—at some precise moment, that I might myself gently rock into a windmill. My legs could become permanent fixtures; my arms could loosen from my shoulders in order to spin freely. The energy coursing through my body could be fed by the passing winds. I could stand in stasis for decades, hissing away birds and taking sight of the hills stretching away, the Cascades prone before me. I could taste on my lips the salt of the Pacific, warm my body with the high eastern desert winds. This, I think, would not be a bad life.

But I've never stopped the car. That long night, driving to the farm from Montana, repeating a journey I'd made so many times before, I prayed. I prayed the scan was wrong, that Sutent had not stopped working, that the cancer was not growing. I drove on. I gazed with eyes full at the passing sentinels, the new markers of a better future, and kept moving, driving, the nose of my car pointed to the farm, my dad, with my internal compass stilled.

I'M WAITING FOR THE ICE MACHINE TO CALVE. I'VE scooped out all the ice and packed it around the salad bar. I need more ice. It's years and years before my mother will die, and I'm waiting.

I've worked in this restaurant—Kiya's—for the last year as the only night waitress. I work Tuesday through Saturday, 4:00 p.m. until closing time. I'm usually out the door and starting the hour-long drive home well after midnight. The restaurant is quiet, situated on the eastern edge of the Puyallup Indian Reservation along the I-5 corridor. Half the building is a smoke shop, where people come to buy their cartons of cigarettes tax-free. The other half is Kiya's, a small restaurant with twenty tables that specializes in nothing in particular and goes through night cooks at a rate of one a month. This isn't a good area. I only make this realization later, much later, years after I learn that the whole place had been shut down for tax evasion and illegal importation and a whole string of not legal shenanigans that I never quite catch.

I am working here simply because it's a job. I spend my mornings in school at the local community college, Pierce College, my afternoons at my other job at the *Pioneer*, a newspaper, and my nights here. I'm hoping that this will pay off soon: I'm tired.

The evening rush of locals and retired folks has died

down. I've already stimulated the salt and pepper shakers, married the ketchups, and now I'm parked in front of the ice machine. My back is to the counter.

I'm waiting.

"Hey. Are you still open?"

I turn, and a woman stands there. She's overweight, a little dumpy, holding a huge purse. Her brown hair is shoulder length. Her lipstick is smeared just off the left side of her mouth. These are the details I remember, even today.

"Yes," I say. "You have the pick of the place. Sit wherever you'd like and I'll bring you a menu."

"What time does the kitchen close?"

"You have about another thirty, forty minutes, so no worries. I'll be right out."

She walks off, avoids the center tables, and slides into the first booth on the right. She picks the side of the booth that faces away from the kitchen. She has her back to me. I don't notice her anymore. I'm staring at the ice machine.

I set the menu and a glass of water down on the table, tell her to take her time, and then walk back through the kitchen and emergency exit into the alley. My line cook, Ken D., is back there sucking down smokes. I tell him we've got a live one but to take his time—she won't order for a few minutes.

Ken is over thirty, overweight, and over it. He doesn't care much about anything. He talks about building a house one day, but hates that the government won't let him. He blasts AC/DC when he cooks and often reminds me that I'm wasting my time with "the whole school thing." He likes to have a beer while he smokes and usually offers me one, followed with, "Ha! I'll turn you in to the cops, jailbait!"

I always grin, not because I think that Ken is the wittiest man I know, but because it's the simplest way to get through a situation I don't care about. I just turned eighteen, but that doesn't matter to Ken, and I don't have the heart to tell him he's mixing up his offenses.

Back at the salad bar, I wipe down the drips and grab a to-go box from the under-shelf. I pack up some of the extra vegetables and a piece of cheesecake. My mom loves cheesecake. I always try to bring her something in the evenings.

Every night after work, I drive the hour back up the mountain to the farm and fall on her bed, exhausted. I stink from the grease of the restaurant, my brain hurts, and I always have a couple of hours of homework to do before the next day's classes.

She likes to nibble on the treat I bring her while I tell her about my day. She'll tell me about her day and catch me up on family news. She and Dad talk on the phone most nights when he is not home, but sometimes the phone lines can't connect in Alaska. My dad doesn't like working in Alaska and being away from our farm, but that's the only way he can earn enough money to support our family. My mom bears up well, but she misses him. My brother is in the Navy, and when he's not out to sea, he calls regularly. My sister is in college and works more, and harder, than I do. But my mother manages us all, and we share the intricacies of our lives with one another.

Sitting on her bed, I find it's one of the few times when we're both tired and we're not cactus-prickly. We're companionable in those moments. She eats, I talk, and then we laugh. Laughter from her is rare; this is before they switched her pain medications around and her amputation quieted down. Those first few years her pain was loud, constant, looming.

I tried to care for her with cheesecake.

I'm sitting in "The Place Where You Go to Listen." It's mid-summer, 2010, and I'm in Fairbanks with National Geographic Student Expeditions, leading an expedition across the state.

The students have filed in noisily and filed back out after about three minutes of contemplation. I don't mind. I sit on one of the wide, white benches placed in the small, white room, and I listen. I'm alone.

I'm here because I've heard of this place. We're only two days in, and have just begun to get to know one another. It's ill-timed, this visit to this place. Instinctively, I know it should come after, after we've seen tundra, boreal forests, Denali, Usibelli, extraction and conservation communities, Talkeetna and Anchorage, poverty, clear-cuts, earth welts and coal seams, broken backs, glaciers and moulins, suburban sprawl, wilderness signposts, caribou, the midnight sun. To be in this room without seeing all of that makes the power of "The Place Where You Go to Listen" fall out of context. David Estrada and I talk through this— we make notes for next year.

John Luther Adams created "The Place Where You Go to Listen" in the Museum of the North on the University of Alaska Fairbanks campus. It's a room that, at its most basic, uses sound and light to translate and interpret the natural world. Seismological, meteorological, and geomagnetic instruments are staged across Alaska and report back in real time to Adams's installation. The readings are translated through electronic sounds that hum and nod and a dizzying array of colors flash, creep, and haunt across glass panels scattered throughout the white room. Dark harmonics indicate cloud cover, gently descending into deeper and deeper tones as day transitions to night. Stars and planets are remembered through sharp twinkling chatters; earthquakes and tremors are translated instantaneously into liquid bass tones; smooth, jazzy nothings murmur of nightly winds scouring the state.

The students wander in and out and I sit, aware of how loud solitude is. A friend came here the previous year and told me it changed the way he saw wilderness. He said it showed him that we can know a place, a thing, best through its elusiveness.

Alaska has been ravaged in recent years. Extraction of goods, of minerals and timber and natural gas and oil and coal and fur and landscapes, has escalated into a frenzy, and industry and infrastructure munch voraciously away at the state. It's becoming harder to find elusiveness, harder to see a black night sky without light pollution, harder to imagine a truly roadless state. The trees are falling and the oil is draining and I believe that, to some extent, this landscape must be aware.

Caribou run crazy the first time they hear thunder and mountains sigh when they feel the tattooing of a road. "The Place Where You Go to Listen" is the first room we can enter and learn to speak the language of the Earth. This is the place where the Alaskan landscape responds, where the harmonics shudder when blasting starts at Red Dog Mine or drilling goes deeper in Kotzebue.

I put the last of the ice around the salad dressings and close the top on the unit. I realize that I've not taken the woman's order and rush out to the tables. She's waiting.

"I'll have the Indian taco," she says.

"Would you like it with real Indians?" I ask, not because I'm trying to be funny, but because it's a line I always say when people order the Indian taco. I'm tired; I'm on automatic pilot.

"What?" she says. She tucks a thick strand of hair behind an ear.

"Do you want sour cream and salsa?" I respond.

"Are they included?"

"You get a small amount with the order, but most people

don't find it enough. You can order an additional side of each for fifty cents."

"Fifty cents for both, or each?"

"Each."

"Um, no. That's another dollar."

"Okay. That will be out in fifteen minutes or so. Are you happy with water, or can I get you something else?"

She orders a Coke, then asks, "Is it always this quiet?"

"Yes," I respond, edging away from the table. "My name is Emily; give a holler if you need something, because I'll be in the back cleaning up for the night."

"Thanks, Emily," she says. "I'll be fine here. My name is Patience."

I walk away. I pour the Coke and deliver it, then head into the back where I start prep for the next day. I'm using a huge, silver chef's knife, two inches thick, and I'm beheading carrots. Chop. Chop.

Years ago, scientists used to speculate that there was a ninth planet out there, a possible companion to our sun. It could be up to four times the size of Jupiter. Called Tyche, the frigid world—if it existed—was believed to be hiding out past Neptune, shrouded in the mystical Oort Cloud.

In 2002, astrophysicists John Matese and Daniel Whitmire, from the University of Louisiana at Lafayette, proposed Tyche's existence.

Matese claimed that the planet, estimated to be about 30,000 times farther away from the sun than Earth, was too cold for visible-light telescopes to identify. Temperatures on the planet, the researchers claimed, would be around negative one hundred degrees Fahrenheit. The silent orb would be so cold that it wouldn't radiate any detectable heat. Plus, its incredible distance from the sun wouldn't allow the planet to reflect any light. Hence, Matese said, it is there, but silent and dark. Waiting.

Other astronomers claim it can't exist because they would have seen it with their telescopes and satellites and radars and computer calculations. Even NASA ruled out the possibility that the planet exists. But Matese speculated that Tyche was out there, lurking. The planet's gravity gave it away.

About a light year away from the sun is the Oort Cloud, an icy solar swamp. This place, which acts as the gravitational boundary for our entire solar system, is the proposed birthing grounds for comets. The tidal fluctuations of the Milky Way and our solar system meet and chafe in the Oort Cloud. For the last two centuries, astronomers have noticed that there is "something" in the cloud disrupting this predictable chafing.

Perhaps that "something" is Tyche.

Matese speculated that invisible Tyche's gravitational pull occasionally causes a comet to dislodge itself from its predictable gravitational trajectory in the Oort Cloud and fly into the inner solar system. For centuries astronomers have seen massive comets soaring by in the heavenly skies, heading for the sun.

It appears that most scientists agree that "something" is hiding in the Oort Cloud, flinging rocks into the heart of our solar system, and whatever "it" is, it must be large enough to heave gravitational disruptions on minor space debris.

Years ago, Matese went on record defending his claim: "Most planetary scientists would not be surprised if the largest undiscovered companion was Neptune-sized or smaller, but a Jupiter-mass object would be a surprise. If the conjecture is indeed true, the important implications would relate to how it got there—touching on the early solar environment—and how it might have affected the subsequent distributions of comets and, to a lesser extent, the known planets."

I like even today the idea of an icy, cold planet floating in the far Arctic regions of our solar system, brooding and bel-ligerent. Every couple of millennia the planet gets so worked up it flings chunks of space rock at the sun. I understand this impulse. After my mother lost her leg, and she sat angry in her wheelchair, I used to stand in the corrals on the tallest corner-post and hurl field rocks at the sun.

If Tyche exists, I envision the planet to be dark blue with black tinges. It wouldn't be showy red like Mars or Venus, nor a gaseous white-orange like Jupiter. Neptune is a pale blue, a wishy-washy, hesitant color that smacks of pastel. Not Tyche. Tyche would be dark and aggressive, the original birthplace of Mordor and ice spirits. Tyche would have a glacial spirit, slowly awakening to anger.

NASA has spent years processing data sent back from the Wide-field Infrared Survey Explorer (WISE) spacecraft. WISE was launched in 2009 on a mission to scan space with infrared wavelengths that should, in theory, be able to detect a cold, broody planet hiding in the icy borderlands. The ver-dict is in, and Tyche has been ruled out of existence.

But I can't let it go. I can't. It's a curiosity that a planet, a world, four times larger than Jupiter, might be circling, watching, waiting, just within the far periphery of our vision. It opens up immense possibility. We've not yet deciphered the language of our solar system. The universe contains mul-titudes. The differences between night and day remain an is-sue of languages, and not of measured light.

My cook, Ken, is stretching the frybread. He's bitching that he isn't native and shouldn't be expected to make this shit. I nod, chopping. He tears his first try and curses. I like it when he swears, because he compounds his "fucks" and "shits" until they blend into single syllables. He roars into the walk-in refrigerator, comes out with round two, and stretches it perfectly.

I nod, thoroughly impressed. Usually it takes him three or four tries. Ken plops the dough into the fryer and starts turning the hamburger. I keep beheading.

I started this job a little over a year ago, when I needed more money for school. I couldn't reasonably take any allowance or financial help from my folks since the car accident, so I picked up this job and another one. With both jobs I could support myself and pay for school.

I slice tomatoes, plate them, and head out front to do a trash cruise. I walk by Patience's table. She's sucked down the Coke, so I do a discrete refill. When I place the glass on the table, she looks up, makes eye contact.

"Emily," she says.

I cringe. It sounds abrasive coming from her, and somewhere in my mind a bell tolls.

"Have we met?" Patience asks.

"I work here most nights," I say, and drift away.

Ken rings the up-bell, even though he could just call my name. He rings it again. I look at him. We're about ten feet apart, separated by a sneeze guard. He jams the bell down again. I smile, by reflex. It is the easiest way to handle him.

I go to the salad bar, slop some sour cream and salsa into a large side bowl, grab the plate, and deliver it to Patience.

"I didn't order that," she says, pointing at the sour cream. "I'm not paying for that. I don't have the—"

"It's on us. My mistake." I back away. I feel clammy. I've seen this woman before. I begin to clean.

I am on my hands and knees, wiping down the espresso-machine fridge, when I hear her call my name. I stand up and she's there by the register. She's looking at me.

"I need to pay," she says. Time has rolled by somehow— enough minutes and seconds to allow her to scarf down a huge Indian taco and for me to forget about her existence.

She's holding her purse in her hands. She sets it on the counter and starts rooting through the things inside.

I nod, tap on the computer screen, and pull up her order. Alarm bells are swinging wildly in my head. I think I know this woman. I start to sweat.

"Thirteen-fifty," I say. I don't look at her anymore. I focus on the right seam of the bag, where it is fraying and pulled.

She hands me her credit card. I don't want to look, but there is no physical force in the universe that could stop me. Her credit card is a black hole, sucking my eyes from the frayed seam, the brown thread, the chip of tiny blue paint stuck, magnetized, to the bag. I want so badly not to look. But this woman's name is Patience, and I recognize her.

The name on her VISA card proves it It's her.

I know, in this moment, that I should hurt this woman.

I sit in "The Place Where You Go to Listen" and hear no shudders. There are occasional, tiny chimes, and the light fades slowly and incrementally brightens red, indicative of spotty sunlight outside. The clamor of my heart and breathing seem to echo wildly, thundering through my skin and transmogrifying into shadow monsters on the back white wall. I avoid eye contact and focus on the spray of pink floating on the glass panel before me.

I want a noise, preferably loud. It is peacefully quiet in this room, indicative, perhaps, of a quiet landscape. But I know frustration, protest, anger, violence, drilling, extracting, using, burning: these things do not go peacefully. They go loudly. They go at the top of their wildfire lungs; they go with rattling chains and out-of-tune guitars; they release themselves as tremendous mudslides defacing continents; they shake with hurricane fury.

Environmental scientist and activist James Lovelock swears this planet is a living organism, and that she (for Lovelock, there is no question of gender) will bat last, ideally with a steel club. She must be aware, must have some idea of what is happening. And if John Luther Adams has, for all intents and purposes, connected an Earthly EKG to her core in "The Place Where You Go to Listen," shouldn't we be able to sit near her hospital bed and witness her wrath?

Why is it so damned quiet?

My mother used to laugh at me as I swelled red and bit my tongue and tried to hold in the biblical flood of angry words that colored my childhood. There is less satisfaction in having the last word, she'd remind me, than in having control and choice of all the words coming out of your mouth. As a child, I battled an effervescent temper, one that my siblings could key up like clockwork. As the youngest sibling, I was their marionette. It took me years to learn to curb my tongue, to rely more on a thoughtful response than on an emotional, heated retort.

In *Winter Music*, a collection of John Luther Adams's essays and journal entries, he writes that "much of Alaska is still filled with silence, and one of the most persuasive arguments for the preservation of the original landscape here may be its spiritual value as a great reservoir of silence."

What I want is noise but, sitting here in this room, I have quiet. I have gentle light, soft chimes, low humming.

Alaska is cagier than me. If she'd followed my direction and howled her protest through dulcet tones, sharp staccato shrieks, and a general bashing and clashing, it would be unbearable. It would be deafening heavy metal in a small, white room, with no earplugs.

But I am afraid of her silence. They say that silence breeds complacency—and such silence might make everyone think that things across the world are all okay. That this is not happening.

Her silence is more subtle, and, therefore, much more powerful.

I cannot think in the middle of a storm. I cannot think when enraged, angry, hopeless. There is a time for rage, and there is a time for quiet. The trouble here is that I don't know which one I need. Which one we might need.

My mind blanks and I become pure reaction, response. But what I know here, in "The Place Where You Go to Listen," is that Alaska demands thought, demands consideration, demands awareness, and she gets this by her silence.

Years later, long after I quit the job at the restaurant and moved on to other things, I used to imagine three things over and over again.

The first: the car crash that had crushed my mother had never happened.

The second: the car crash that had crushed my mother had never happened because Patience hadn't left her home that day.

The third: two years after the collision, when Patience had come into the restaurant where I worked, and had eaten dinner and paid with her credit card, and I'd looked her in the face, that I'd grabbed her face and jammed it down into the glass case. I imagined pouring boiling water on her flesh and making her feel for one instant the level of pain she had caused my mother. It's the type of pain I cannot fully fathom, but my mother dealt with it daily for the rest of her life.

I'm aware that we're not supposed to talk about the details of our dark imaginings. But these were mine for the longest time. I imagined what pain I could cause this woman. I never imagined yelling brutal words or dramatically identifying myself and watching her cringe. I just wanted to physically hurt her. Make her understand what I saw my mother negotiate daily.

I don't imagine that revenge anymore. I used to carry it around, a heavy video playing in my pink brain, unable to switch off. I used to imagine that scene rather than focus on what had actually happened. I used to imagine that scene because it made me feel not powerless. In the face of the enormity of everything that had happened to my family, it felt like one meaningful act. No one told me what to do; no one gave me instructions. I saw the mess, and I didn't know what to do.

This is not a story of forgiveness. I don't forgive Patience. Her name is emblazoned in my mind, tattooed inside my eyes. But never is my first thought in the morning of her, and my last thought of the day is not of her. I think rarely of her, if at all.

It was the sick humor of the universe that we crossed paths again, that I recognized her. I'd never imagined it happening. But then, so much of what happened in my life, to my family, I had never imagined. Writer and activist Rebecca Solnit reminds us that the future is not dark because it is terrible; rather, it is dark because it is yet unknown. Our real future is unimagined, and all the small twists of fate are inconceivable until they actually occur.

The unknown does not terrify me. I experienced unfathomable terribleness, and I survived. I look ahead into an impossibly dark future, and I see the bright lights of the possible.

I never wanted to really hurt Patience all those years ago. That first moment when my system shocked itself into recognition—all I wanted to do then was react.

I didn't attack Patience.

I didn't do anything.

I handed her the credit card. I waited for the machine to spew its tiny slip of paper, and I passed it to her to sign. She didn't have a pen, so I pulled one from my apron and gave it to her. As she signed her name on the bottom of the slip, I turned and stepped into the back of the restaurant and into the walk-in freezer. I stood there, in the cold dark, and I screamed. Just once. In my memory, there is debate over whether the noise that came out of my mouth was actually a scream, or if it was more like a yelp, a bark. But that's it. I opened my mouth and emitted a noise. That was my single action.

No one heard. Patience did not hear. I imagine she left the restaurant, went home, and lived her life. I doubt she had any idea who I was, or of the torture she had caused me that night in the restaurant. No doubt, no doubt, she was clueless as to the boiling insanity she momentarily brushed up against.

I left the walk-in, came out front, and picked up the slip off the counter. No tip.

I remember going back by the salad bar and picking up the restaurant phone and calling the farm. My dad answered, home from Alaska, and we talked briefly. I recall telling him that Patience had just been in the diner. I asked him what I should do, if it would be of use for me to run after her and kill her.

My dad, of course, counseled against that course of action. I wept then, on the phone, listening to his voice. I don't remember the rest of his words, just that he spoke to me and

I cried, and Ken revolved around me, mopping the kitchen floor and cursing to himself.

After hanging up the phone, I didn't finish my job. I took all the money and credit card slips out of my apron, bundled them up, and tucked them into the safe. I grabbed my keys, walked outside and to my car, and then drove home. The next evening, citing sudden illness, I apologized to the afternoon waitress, who had screamed at me.

After going home that night, I lay across my parents' bed without cheesecake or treats, and I cried. I remember my mother telling me that inaction was well within the realms of grace, that I did my family and myself a great service by refraining from violence. My dad advised me to take the anger, the hurt, the sadness, and the rage, and package them together and consider sending them out into the night sky, past the moon; I should will that ball of emotional distress far out into the galaxy. He told me to let it cool there for a millennium or two, and come back to it when I was ready.

I don't believe it matters whether Tyche actually exists. Is it not enough to imagine such a world might be out there? Can I not revel in the wonder of possibility, in the hope that, for centuries, what was known about our solar system was not, in fact, totally known? That our realm of possibility is endless? That there is so much unknown, so much more to be found?

I sit in "The Place You Go to Listen," and I know what I want to hear. Thankfully, I don't hear it. Instead I hear what is actually there. I remember a low hum, continuous, in the background, low. I know the sound was there the entire time, but I didn't hear it until I was listening.

The namesake of Tyche is the Greek goddess of fortune. She hands out both good and ill will, depending on her moods and the fickle currents of the winds. Ancient pictures of Tyche depict her holding a rudder in her hands.

It is possible, I tell myself on late evenings when I wander the Alaskan backcountry, that astronomers have just now found the anger I sent out into the night sky all those years ago, the rocks I chucked at the moon in frustration and rage. Maybe they left our atmosphere, became truly free, attached themselves to galactic dust, and disappeared from my life. Perhaps, as the night sky loomed palpably close over tundra and frost, my young anger morphed into an icy ball called Tyche.

I have no idea what isn't possible at this point.

A COUPLE OF WEEKS AFTER I FELL INTO THAT GLACIAL POOL and dreamed fish dreams, I was back out on the Meade Glacier. At the time, before my mother died, I worked as a guide on the glaciers surrounding Skagway. People would helicopter out to the ice, and I'd be there to greet them, explain the area, and ensure they didn't slip or slide deep into the ice.

I adored being out on the ice. I'd take a wander, walk across bare swathes of ice, over moraines and crevasses—eventually finding a place to perch, to look and watch and listen and browse my thoughts.

There is a pervasive sense of kinsmanship that forms on a glacier. There are places to grow, to walk, to exclaim over. But in those moments, deep in the soul is a sense of recognition, a sense of having been here before. If it is your first time out, the newness isn't as new as you thought it would be. There is something subterranean about it all.

It is impossible to know what really runs beneath the surface of each person. We could each share our real selves if we chose. To be known is to invite others into your river to course and pulse with the waves of your life. But a glacier: this is different. It is entirely possible to peer deep down into that subterranean river. The coarse surface of a glacier is riddled with insights.

I remember my first days out on the ice, exploring and exhilarated by the spaciousness of it all. The Meade Glacier was miles and miles long, miles wide, and hundreds of feet thick. You could range out in any direction and keep going, stepping and stepping, crunching your crampons into frozen oblivion. I reveled in this and spent the free hours soaking in the glacier's details.

Glaciers are alive, with brittle, scaly skin. Most places, if you stop long enough and stare at the ice around your feet, you'll see what are best described as glacial pores. They look like circular holes of varying depth, with black algae goop piled at the bottom—cryoconite.

Cryoconite is a primordial soup of dust and microbes and soot, individual ecosystems in glacial Petri dishes. Imagine first your hand, and then place a big, black freckle right in the middle of your palm. Hold the palm up to the sun. Feel the freckle get warm. If your arm were a glacier, all the solar radiation, the warmth, that the freckle would absorb would heat the ice underneath it, and slowly—in a day, a glacial moment—you'd see the freckle sink down into the ice. The process of sinking down carves deep, cylindrical holes. And as more ice melts, water gathers, marooned in the hole, swirling around with the cryoconite.

Originally just bits of dust, flecks of soil blown in from the mountains, the cryoconite attracts organisms to take up residence: algae, insects, copepods, ice worms, even tardigrades. Cheap rent is advertised in the glacier newspapers, tempting offers of free furniture and pleasant neighbors.

On the surface of a glacier, first glances show not much to be alive, thriving, growing. But stop, bend at the knees, and you'll see that the ice is a metropolis of small cryoconite holes splattered indiscriminately. The eye will find patterns—perhaps make patterns—among the holes. Once, I saw a patch of cryoconite holes that resembled a reverse-Braille pumpkin speaking in secret pumpkin language, available for anyone or thing that learns best through pumpkin talk. Absurdity in patterns.

Scientists have suggested that cryoconite holes are communities that have perceived boundaries with specific energy flows and nutrient processing. The residents of each community exist in isolation from one another, possibly. I'm not sure. I'd bet they know of the surrounding communities. They don't think they're alone in this world, burdened with having to figure it all out for themselves. Instead, they communicate with their neighbors, pass back and forth hints for glacial living, help each other out as the ice melts. They might even, I think, have town councils, neighborhood watches, perhaps even a United Nations of Cryoconite Communities.

Hiking in the forests that fold down around the Laughton Glacier, which feeds off the same ice field as the Meade Glacier, makes me a believer in magic, in fun, in absurdity. Technically, the trees are part of the Tongass National Forest— an enormous rainforest. Here, the big trees growing on the edge of a continent are patchy new and old growth. Together, the rainforest contains more biomass per acre than any other place on earth. To me, it is a sun forest, a place where ferns have tongues wrapped around their necks, where beams of light tell histories free from accidents and blame, where it becomes infinitely clear that one of the biggest differences between humans and trees is simply that humans burn trees.

Watching the birds drink water drops off lichen webs and press their faces under their wings makes me believe this should be holy place, a place to go to remember and move. By "move," I mean "express energy."

It is possible to come to this forest and not proselytize about the mysteries of nature, to not be tipped spiritual in a matter of seconds. But then, that's rare. It is hard to over-look the rawness of this place, to not wonder at the glacier that sits high above the forest, intertwined with the jagged peaks. And to wonder at the fact that the glacier has freckles dotting its surface, freckles that heat and melt down the gla-cier and aid in its ability to channel its inner soul. Consider that what was once an unquantifiable amount of snowflakes slowly morphed into glacial ice, and then was pressed from ice into water by cryoconites. And that water rushed through the glacier, pure and pervasive and clear, and pushed out be-low to traverse the bare landscape and seek out forest and fauna. The surrounding old and new growth, the trees and shrubs and lichen and mosses, most growing in less than an inch or two of soil—these beings pull that water up through their roots, they surge and pull and quench themselves full on uncorrupted glacial water.

And you can't see that while standing in the forest; you can't see the hum of busy water running wild underfoot and in the air. You can't yet imagine the ice overhead. But if you stand still, if you just pause for one moment and get away from your busy mind, you'll feel it. The subterranean pulse.

In my life I have stood upon many a mountain ridge and felt intuitively that I had been there before, that this was not the first time I had stood on a razored edge of earth and had gazed down into the green of life below. There is evidence of this everywhere I look: this pulse, this energy. I don't know exactly what this energy is, but I rec-ognize it when I detect it.

I think that it is, perhaps, a planetary imagining. Is it pos-sible for a planet to imagine?

I remember once falling asleep in the midafternoon on the Meade Glacier. The helicopters had come and gone and it was silent and I was curled into the blue camp chair outside the tent. There weren't any people, and all the other guides, Mario, Kyle, and Jason, had wandered off. I luxuriated in layered warmth, all my limbs folded together. The sun shone down, making the ice hazy. I nodded off and awoke much later without even realizing I'd fallen asleep.

The dream stays with me today. I dreamt that there were ice spirits, corporeal embodiments of every single glacier in the world, and that they had gathered together on the cusp of their deaths. They spoke to one another, and their voices sounded like ice calving, thunderous swishes accentuated by melting tones. Like they were cold relatives of Treebeard. They argued, then, about existence. To go or to stay. And there were no colors in my dream, but then, today, somehow, I remember that each glacier had blue eyes.

I dreamt this meeting, and I dreamt that I was there; I was watching. And as they filed out, away, beards crusted with boulders and lichens, one looked at me and promised magic.

Glaciers are keepers of magic, of imagination. What will happen to the magic when they are gone?

What if there are cryoconite holes in the fabric of our world? What if we're able to peer down through them and not see small communities of tardigrades, but instead see the ineffable currents of magic and imagination and energy that weave throughout the tides of our existence?

This is not a question of "what if" for me. I have looked down into these rabbit holes, and I have found wonder.

After my mother died, I looked everywhere for a note from her. A sign. I had no faith that my life would continue, that I could move ahead without a mother.

I looked everywhere, and I found nothing. So I returned to Missoula, went to graduate school, and spent my free time telling myself stories. But even then, walking the city streets, lingering by the river, following the smell of my mother's gardenia perfume on a woman in the grocery store—still I looked.

And then, unexpectedly, I found it.

A minute, penciled check in the dictionary. Lightly drawn. V-shaped, with the right arm of the *V* much taller. As soon as I saw it, I felt a physical staccato, like a freight train, rumbling across packing peanuts, rattle up my spine. Laying eyes on it just weeks after her death sent me deep into the core of the earth, so deep that not even the most shining of life rafts could enter. A decennial increment lasting longer than a glacial lifetime passed without my notice.

I had taken the dictionary with me when I left the farm, after we had spread her ashes. I took the dictionary because I honestly didn't know what else to take. I wanted something I could have in my house that was familiar and real and sturdy and tangible and hers. I took the dictionary.

I made it ride shotgun the entire way back to Missoula. I swear it looked at me.

It's the *New Oxford American Dictionary*, published in 2001. She gave it to me in 2002, for Christmas. It's from her and Dad. She wrote on the first page, "To Emily, Love from Mama and Dad. Christmas 2002."

The dictionary weighs over ten pounds. It is eleven inches long by eight and a half inches wide. She gave it to me, but I did not take it. Not then. It never left the house, the farm. It rested, presided over the center of the living room, and was a fulcrum for the enormous floor-to-ceiling bookcase that stretched up, down, back. Its position implied it was a key to the thousands of books scattered throughout our home. She had used it regularly; we all had.

There are little gray check marks cast throughout the dictionary. Every time my mother looked up a word, she'd mark it. Some words have several check marks by them, implying she looked up the word several times. Like the word "imprimatur." Noun. Meaning an official license by the

Roman Catholic Church to print an ecclesiastical or religious book.

The game I play with myself now is to figure out what book she was reading when she came upon a word she didn't know. We've read many of the same books, and often, I look up the same words. "Peripatetic." "Oubliette." "Otiose." "Oubliette" and "otiose" have the same check, the same color and weight, different from "peripatetic's" rather lavish check, a check that stretches right up into "periostitis." Perhaps whatever she was reading had the words "oubliette" and "otiose" in the same paragraph, or in the same sentence. And I wonder then, what book was she reading that had a French secret dungeon with access only through a trapdoor in its ceiling, which may possibly have been serving no practical purpose or result?

This book, this game, this word-finding—it is my imagination, my rabbit hole. My mother and I—today we talk through ambiguous, cloudy symbols dotted throughout a dictionary. These marks lead me to books she might have read where, in all likelihood, she'd have written a comment or made a mark. I search for her penciled comments in the margins, see what she underlined, checked, boxed, starred, exclamation-pointed.

It is a way to know her, to understand her, to see what intrigued her. To think about the words she marked a "Yes!" next to. As much as this sheds light on what she liked, it also increases the mystery. Why did a sentence about the unscrupulousness of children get a check-plus? Why did the word "morose" get underlined three times? I am weighed down now by so much I don't know, so many stories she said she'd talk about later, so much that went on between us as I struggled vainly to separate her disease from her. One, I hated; one, I did not. Sometimes, I think it might have appeared unclear.

Late in the evenings, when I am tired and my walls are weakened and I am in need of magic, I approach the

dictionary like a supplicant. I let my mind wander avenues that a more vigilant me would not allow. I demand
the dictionary to tell me its secrets, to replay conversations
she surely must have had with it on her own late-night,
weary quests for meaning and nuance. What a choice it is,
sometimes, to get up from where you're ensconced, traipse
across the wooden floor, iron nail heads poking into toes
made sensitive from a full day's worth of stepping, flip on
the light, and then turn, one by one, the wispy pages of
the dictionary. To squint at the tiny print and hope to find
your one word out of over two hundred fifty thousand
definitions. Each time, there is a choice to be made.

Did she ever read a book and find a word about which
she wasn't fully clear on the meaning but then choose to
overlook it and move on, gathering a general definition from
its usage and the surrounding words? Or rather, would she
be so moved by the word's sound, its beauty, that she would
heave herself up from her chair, strap on her leg, and walk
across the room to satisfy her curiosity? Did she murmur as
she walked, repeating the word, "Oubliette, oubliette, oubliette," sounding it out, saying the *oo*-sound or the *ob*-noise,
the harsher, more finite version tumbling over the softer coo?

I know the dictionary can't tell me these things. But still,
I ask. I have placed it on my table, within reach of my lap-
top, surrounded by photograph frames. I circle it, I look at
it, I demand something of it. I flip random pages just to find
marks, checks, dots.

We have a staring match, this book and I. I flip through
the pages and recognize words. This is a dance I do so many
nights, watching the darkness cast long shadows across the
pale dictionary pages. I am trying to communicate with her,
and it is working, even as I am aware that this type of wishing
is an invasive species to the soul. The dictionary is not magic
itself; rather, it is the rabbit hole I choose to plunge down
when I most need to dream.

In 2010, when I was at Chena Hot Springs for National Geographic, we studied the heating capabilities Chena demonstrated through the utilization of their hot springs. The energy production capacity of Chena's system is immense. There is a primal earthiness to geothermal energy. It is a comforting energy, a feeling of reconnection. There is a sense that this has been done before and that there is power running along below our feet.

Geothermal means exactly what it sounds like: "geo-" signifies earth; "thermal" means hot, hot, hot. Earth heat. It is the oldest of energies on this planet, originating deep within when Earth was first formed. Volcanoes and solar energy absorbed by the planet also create geothermal energy, but most radiates out from the hot molten core, the heart of our planet.

This geothermal energy manifests itself most commonly in the form of sulfuric hot springs scattered across the surface of the planet. We go to hot springs to bathe in the heat of the earth, cocooning ourselves in steaming ponds like we're snuggling up to a lover in bed.

Mark, our facilities guide at Chena, spoke at length about the sheer accessibility of geothermal energy. Previously, it was thought that geothermal energy could only be accessed near tectonic plate boundaries. But technology has expanded, allowing for geothermal energy to be available virtually anywhere.

"Men used to deliver everything on horses," Mark told us. "When oil came onto the scene, the horse industry was resistant. Oil is now the industry and resistant to anything else. They're against geothermal, but this will be our new future. We can even use old oil infrastructure to make geothermal energy available on a massive scale."

Mark was referring to depleted oil wells that traditionally

have been filled with water and capped. These wells are generally deep enough to create enough geothermal energy to heat small-scale projects such as homes, greenhouses, or small townships. Small demonstration projects have highlighted the success of this reuse of the wells. Oil companies themselves are exploring this technology to investigate whether they can offset various electrical costs by generating geothermal electricity on-site.

"The biggest hurdle to geothermal energy," Mark said, "is cost. There is a multiyear ROI (return on investment) that often stalls projects. It is expensive to begin, but once going, it goes."

We discussed this among ourselves as we continued touring the Chena facility. When is the price too high for renewable technologies? How is that price measured? If human societies can live on this planet, harvesting geothermal energies with minimal negative impacts, is there really a price that is too high?

Later, after we'd visited the geothermally heated cabins, greenhouses, and various other buildings, we dipped into the sulfur waters to relax.

One of my students mentioned how efficient it seemed to reuse the depleted oil wells. If they're already there, why not use them today?

Oil wells provide us with a look beneath us, deep into the earth. They are our cryoconites, our telescopes into the magic. There is energy below us, hot, molten energy, and we could run a world cleanly on it. Yes, Mark is right. It is an expensive hole to climb down into, but as Alice discovered, it is worth every second. As a nation, we can live without geothermal renewable energy. But should we? Since 1943, Iceland has been heating its modern homes through geothermal energy. Hot water is piped to heat each building in entire communities, and the pipes themselves are routed under sidewalks so they can melt the snow and ice above to clear the pathways. In actuality, geothermal energy has been keeping Icelanders warm since the time of the first human settlement on the island in the ninth century. Evidence of this relationship between Icelanders and geothermal sources is everywhere. For the last five summers, I've worked in Iceland, and what amazes me continually is that most Icelandic towns and villages have their own community hot pools in which to bathe, commune, and relax.

It would be only a small stretch to begin doing this on a wide scale in the United States.

The costs are comparative. Continuing business as usual and spending enormous amounts of money on non-renewable resources is proving to be disastrous for our planet. Compare that with the cost of investing in geothermal infrastructure and utilizing a widely available resource that does not fuel climate change. Choices like these are startlingly clear.

Riddled across the planet are hot springs, small, gray check marks that remind us that there is energy all around us, magic just below the surface. What we do with these checks, how we start the conversation, determines how we want to steer our future. Don't we hold within ourselves deep geothermal inclinations, hot furnaces burning in our chests that pull us and connect us and touch us? We may seek that heat out in others and find it in the most surprising places on Earth.

Here on Earth there exist cold and heat, magic rabbit holes peering down into the cold blue-blue of ice and the dark black of molten cores.

I used to run transit with the dictionary for my mother.

She'd be in her bed, or on the couch, or on the outside

decks that surrounded our house. She'd be reading a book and she'd discover a word that she didn't know, and she'd query it, highlight it, puzzle over it. If I was around, she'd call to me, direct me to the dictionary.

Our house is small, and her voice would echo off the walls while she called letters to me bingo-style. The pages would fly as I'd search, finger sliding down the rows.

Once, we sat outside with coffees in the early spring, watching the hummingbirds war with each other around the hanging feeders. She sat on a bench Dad had made, leaning back against the shady cedar wall of the house, her oxygen tubing snaking down from her face, through the open door and the house to the tank upstairs. I sat to her right, perched in a lawn chair, with my feet resting on the chopping block for making kindling.

Everyone in our family looked forward to the coming of spring and the arrival of the hummingbirds. We called them "hummers" and surveyed the skies extensively in the early spring. The males always came first, zipping around after the first cherry blossoms. It was quite the occasion to report back those first sightings. Mom was often the one who claimed the "first spot."

First arrivals were heralded by bright red flashes from the trees and shrubs as the males bounced sunlight off their chests in the hopes of luring a lady. I remember specifically this day, because Mom and I sat watching them, trying to identify which male claimed which half of the deck and watching the aerial feats he performed to cement this ownership.

We laughed then, a lot.

We were reading about hummingbirds from a guide-book resting on Mom's lap. I kept running to the dictionary, stepping over her oxygen tubing as my mother spelled words she didn't know, and inside the house, knees bent as if at a Puritan hearth, in front of the dictionary, I would mine for clarity. We'd stumbled upon "nictitating membrane." A third eyelid, capable of being drawn unconsciously across the eyeball of a hummingbird for protection. A built-in set of goggles. Personal, protective hummingbird equipment.

Mom's *National Audubon Society Field Guide to North American Birds* was a trampoline for vaulting towards tantalizing bits of hummingbird information. Hummingbirds are solitary, aloof, hermetic, resentful of the forced socialization they must endure as they sip elegantly from a communal feeder.

We unraveled these facts piecemeal, enjoying the individual mystery. Once the third eyelid was identified, we were off, reveling in ossicles and spiderwebs.

That day, we volleyed information back and forth. From my guidebook, I told her that hummingbirds have impeccable memories, able to remember exact flower locations and maintain a careful monitoring pattern to maximize nectar refill in the plants. I demonstrated by moving our hanging deck flowerpots: havoc ensued among the hummingbirds, who'd created regular patterns around the zinnias. But they were quick to home in on the new nectar spots: they have eyes larger than their brains and possess razor-sharp eyesight.

She told me the female hummingbird is solely responsible for building her nest and raising her children. The hummingbird will construct her nest primarily of small pieces of lichen, woven together with spiderwebs she's poached from dark, dim corners and painstakingly delivered strand by strand, trip by trip, flight by flight. She'll take elaborate measures to camouflage her home. She'll chip paint off the surface of the building and decorate the outside of her nest with the paint specks to better camouflage it.

Mama hummingbirds can be fastidious little creatures, often seen contemplating their homes in different lights. Based on where the light falls, they'll carefully place lighter-colored pieces of moss or lichen and then blend them into darker pieces of debris. Every piece is methodically placed. It must be.

When hummingbird babies are born, they weigh about a third the weight of a dime. The mother will lay her eggs at separate times, but through hummingbird magic and imagination, she makes all her eggs hatch at once, instantly filling her nest with tiny, hungry babies whose number-one job is to remain silent. To keep them quiet, she feeds them around the clock, regurgitating sugary nectar and protein down their throats. Her *W*-grooved tongue will hold their baby tongues down as she passes the nutrients.

I remember all these facts. I remember the conversation my mother and I had as we sat on the deck and watched the hummers buzz the feeders, marveling at the tiny creatures. I remember the silence we shared as we listened, hot cups of coffee in hand, warm despite the early spring chill, watching.

I want to revisit those moments again. I have new information. I've read more. I've studied. But what do I do with these new facts?

Now, I could tell her that hummingbird sex lasts approximately four seconds, that a male hummingbird doesn't have a penis. Wouldn't she laugh at that?

I'd tell her that a female hummingbird is born with two ovaries but loses her right ovary within her first six weeks of life, probably, ornithologists speculate, to lighten her load and make her quicker. That after mating, a mama hummingbird may never again see the father of her children. If he ever comes by for a visit, she'll chase him off, fearing that his bright colors, which so attracted her in the first place, might signal predators to come investigate her nest. She'd rather be a single mother than risk her children getting devoured by monsters.

But my mother isn't here to tell, so I direct my comments at the dictionary. But the dictionary never responds. It is immobile. I know this, and I am frustrated because she is not here to talk to. She is not responding.

Why is it that nothing is responding to what I am trying to do?

A mama hummingbird would rather be single than risk her children. Did my mom ever think about what it would be like to raise her children alone? Dad worked at least half of each year in Alaska, making money and sending it home. When he was gone, my mother raised us and tended the farm and did everything that had to be done. She didn't shirk, rest.

I think a lot now about the questions I never asked her. How did she manage to walk away from her previous marriage into the arms of a man she barely knew—my father—and successfully build a life with him that spanned the next forty years? How could she have taken such a risk? Why can't I?

We used to sit together outside on the decks and marvel at the world. We shared information, books, thoughts. But what I need to know today I did not think to ask then. And now, there are no instructions. No guidance. I am moving forward as best I can.

My parents met in the 1960s in California, both full-blown hippies. Both carried the heavy burdens of difficult childhoods and alcoholic parents.

My dad's father died of cancer in Detroit when Dad was fifteen, after several years of illness. A couple of months later, the family packed up what was left and moved out west. My father drove the family car, pulling a horse trailer, across the country. He used to tell us about that journey:

"Your grandma, too drunk to drive. So I drove, and Lyn drove the other car behind me, barely peering over the steering wheel."

My Aunt Lyn was thirteen years old.

They landed in Arizona to stay with relatives for a while, but then their fortunes seemed like they'd be better farther west, so they packed again and moved to California. Grandma drank and married. Dad folded himself into California hippie life with lots of booze, drugs, music, and surfing. He used to say that he found peace in the big ocean blue. He surfed so much that he became quite skilled: producers used footage of him surfing to open the film *Endless Summer*. My dad used to brag that he could roll a joint up his arm while surfing. This statement would always be delivered with a twinkle in his eye that barely hid his nostalgia.

Mom grew up in Bakersfield, California, under the dark shadow of her mother's mental illness. Her father divorced her mother and left; he remarried, and for a couple of weeks each summer, my mother's childhood found peace when she got to stay with her father. But life at home was turbulent, and she left as soon as she could. Her brother, my uncle Larry, went to Vietnam, and my mother struggled alone. She had two marriages under her belt by her mid-twenties. Rarely did she speak about that time in her life: clearly, it was unhappy and violent.

One evening, going out to buy pot with a friend, she met a tall man with shaggy, blond hair. A couple of days later, they happened to be on the same train, heading north to visit friends. By the time they both stepped off the train, they were in love.

There is a picture of the two of them from around that time. My dad has a large, scraggy, blond beard and long hair. He's looking at the camera from his right eye. Oddly, he's wearing a button-down shirt, which seems out of place. Mom looks young and slightly haunted in the picture. She's snuggled up to Dad, tiny-framed in comparison to his six feet and two inches. Her hair is long, straight, but that's not what dominates the photo. It is her brown eyes, which look directly at the camera. There is a hint of a smile on her face, but that smile doesn't fully reach her eyes. They hold an expression of waiting, as if holding back because soon the other shoe would drop. Undoubtedly, she was still married to another man, not Dad, when the picture was taken.

They shared the same dream, to build a home and tend a family and each other, and they shared the same birthday,

May 17. Dad was two years older, but they took it as a sign from the universe that this—this thing between them, this love—was meant to be. They married in Grass Valley, California, in a celebration that Mom's family didn't attend; neither did much of Dad's. No one believed it would last. In fairness, perhaps my parents weren't so sure either. But after that, they never looked back.

They headed north after they married. My dad, telling the story:

"We had eighty-five bucks in our pocket and knew we had to go to Canada to homestead. So we took off."

They drove Dad's faded, red 1953 GMC flatbed truck. Dad had built a small house on the back, and they had their dogs and cat and books and each other. Driving north took them to Slave Lake, where they briefly settled. My Aunt Lyn joined them; they built a community, friends. They found happiness. Lyn took to Canada so much that she never left, eventually settling on Vancouver Island and becoming a Canadian citizen.

"We broke down somewhere on the ALCAN. I remember a guy stopping to help us. We had a broken axle, which he fixed by using a wire clothing hanger to spot-weld the patch up. We limped into the nearest town, but hey, we got there!"

Later, the winter in the north got to them and they moved south, travelling again on the ALCAN, the Alaska-Canadian Highway, landing at last at a small cottage built on stilts at Salmon Beach in Washington State. There, Mom tended bar at night while Dad went to school and pirated drift logs from Puget Sound. My dad always claimed he was "liberating" those logs. In my mind, I see them in the hazy, late-evening light, cruising the inlets of Puget Sound in a small boat flying a pirate flag.

Years later, in the mid-1970s, they decided to have children and went farm hunting in the foothills of Mt. Rainier.

They bought our small farm, named it The Rainbow, and inadvertently became part of the wider movement of their generation to be back-to-the-landers. Own your own land; grow just what you need; live simply. Life worked then: they both stopped drinking. My mother would be involved with Alcoholics Anonymous for the rest of her life, helping other people reach sobriety. They raised three children. I was the youngest.

Moving the tent on Meade Glacier was a ritual my co-workers and I all loved and loathed. We once moved it twice in two weeks: a record.

We had a tent set up, a home base to shelter our gear and us when we weren't showing folks around on the ice. It was large, white, expedition-style, with walls that could flap up into doors in minutes. It wasn't the biggest, and Mario and Kyle had to duck their heads significantly to enter and exit the tent. For once in my life, my shorter stature was a boon; it made tent access an act of ease. The tent didn't have a floor, and it was rooted to the ice by rocks we'd dragged in from the moraine and placed around the outside skirt.

We would move the tent when even I couldn't get inside anymore.

Imagine a large umbrella parked over a section of glacier. Then add a hot, yawning ball of sun, shining down on that glacier. Ice melts quickly. But not under the umbrella. Soon the entire surface of the glacier is a foot or two shorter than the sheltered ice.

That's what happened to our tent. The actual surface of the glacier was melting, deflating. After a couple of days, we'd be stepping up into the tent, and the tallest of the five of us guides, Mario, soon wouldn't be able to stand up at all inside.

So we would get together, unhitch the guide lines, move the rocks, and then, collectively, pick up the tent and shuffle it left, or right, or down-glacier a hundred feet or so. Then we'd go back up, collect our gear, and shuttle back and forth a couple of times to complete the job. Tie the tent down, roll the rocks back onto the skirt, and call it a day. We'd collapse into the camp chairs, throw our feet up off the ice onto the gearboxes, and rest. Someone would start to whistle.

After several weeks in a given area, the surface of the ice began to look like it was pockmarked with crop circles made of ice. They stood out, oddities, the only real temperature gauge for how quickly the entirety of the glacier was melting. It is common to assume that glaciers only recede or surge at their toes. But their actual masses will grow or shrink, given circumstances like snow and temperature variables. It seemed, that summer, working out on the ice, that the glacier was shrinking rapidly, a foot or two a week.

In that second week, when we were taking the tent down again, I remember standing on the backside of the tent, lifting a rock, when I caught a beam of red out of the corner of my eye. I ratcheted my head to the left, quickly, to see what had caused it. Color on a glacier is out of place. You can see off to distant ridges the browns and greens, but on the ice itself, there is little color. White and blue. Ice and sky. You don't notice the absence of color until you see a blast of iridescence, a feeling of energy, a hue of magic passing over your eyes.

I looked for the source of red and couldn't find it. Sometimes, out on the ice, I saw things that weren't there, or sensed the breezing of glacial angels. I gave up on the red, put it out of my mind, and went back to dismantling the tent.

But then, from the other side of the tent, I heard Elizabeth Ruff, the only other woman out on the glacier with me, shriek. I rushed over. Elizabeth is a woman given to shrieking, but generally what is worthy of a shriek to her is wonderful to me.

"A hummingbird! Did you see that?" She looked at me, eyes wide.

I shook my head.

"It just flew by my face!"

"Are you sure?" I asked. "Really?"

"YES!" Elizabeth gave another shriek.

We stood there, stock still, eyes wide and waiting. It took a minute, but then a zip of green slipped between us. It circled Elizabeth's head, the body and buzz clearly identifiable.

Mario walked over. The bird buzzed him, zipping in, out, away. He laughed, a huge smile lighting up his face. I hadn't seen him smile like that all summer.

I was astonished. We were out just below what was called "the Y section" on the Meade Glacier, an area where two glaciers merged into one and fed down into the valley. This wide area sprawled over a mile from valley wall to valley wall. The toe was several miles down, and there were miles and miles of the Juneau Icefield to the south. I couldn't imagine that this tiny little bird had flown all the way out here. Occasionally, we'd see eagles riding thermals above the ice, but rarely any other bird. And, just then, in front of our faces, a hummingbird!

Magic!

Elizabeth, Mario, and I wore bright yellow guide jackets. Was it possible that this hummingbird had energized himself to fly all the way out there, to the deepest maw of a glacial desert, just to investigate the possibility of our being the biggest bonanza of nectar in southeast Alaska? Did the bird mistake us for a flash of heat, a phenomenal, glacial, thermal energy?

I don't know. But I was in wonder at its presence, concerned for its return journey, and grateful that Elizabeth and Mario were there to witness something I was not positive was real. Mario and I had been arguing before the humming-bird visit. Afterwards, we didn't argue.

Imagine if it were a spirit, a small glacial wish that surged by on its way to fruition. Perhaps glaciers imagine in the form of hummingbirds. If I were out on the ice today, and that happened, I would imagine that it was an expression of love from my mother, a sign that she knew I would surely recognize, one that didn't demand nightly Ouija rituals with the dictionary. The memory of the hummingbird is like earth heat to me, a verifiable fingerprint of greater forces at work.

I'm on the phone with Jordan Vroom, a good friend of mine who lives on a small island in Washington State. It's a year and a half since Mom died, and I'm out at the farm on a long weekend home from graduate school. I'm sitting on the tile floor in the bathroom, trying to figure out my dad's prescription medicines. Jordan and I have been friends for several years. We met in Zambia, during our Peace Corps service; I attended her wedding to Thomas and held her confidences and worries, and we've wandered across each other's farms and dreams for years. Jordan's a calm woman with hippie tendencies, brown hair that is sometimes long and sometimes short, a woman with flat feet and a tinkling laugh and jewelry often accessorized with birds. She's listened to my musings more times than I can remember.

On the phone now, she's building up to something; I can feel it. Our conversation is circular. She's asked me how I am doing several times. I've answered. We're in redundant land. Finally she just says it.

"I want you to be my baby's godmother."

Jordan is three months pregnant with her first child, Cooper. This is what she's imagined, longed for. Her husband, Thomas, is a long-haired electrician for a solar panel company in Seattle. They live in a small house and raise chickens. There is talk of getting a dog soon. Jordan wants a family, and this baby inside of her is a faithful step in that direction.

I don't know what to say.

Jordan continues: she wants me in this role, but she has concerns. She's worried about the future, my future. "With your Dad sick," she says, "where will you go? After? What if you leave the state? How will you be a godmother from far away?"

I pause, reach for words.

"First, I am honored." I say this to her because I am going to say no. I am going to refuse this friend, this woman who held me when my heart broke, who chased away snakes from my hut in Zambia, who came out and helped my dad rototill the garden when I couldn't be there.

"I think your concerns are valid. I don't know where I am going to end up when this is all over. I don't know even if I am capable of surviving this. I can't imagine what is next."

There is silence on the phone. Jordan understands.

"I think I am a mossy Northwest woman. I think I will end up here. But I don't know. I have no idea what the future holds. It is so fucking dark."

I tell her that my hope is to go to the farm and take care of my father until he dies. And then, optimistically, through the fortunes of fate and a heavy dose of miracle, I might be able to stay on the farm. I could live quietly and write books and have a peaceful life alone.

I tell Jordan these things, but they are half-truths, and I am aware of this. Jordan wants me to promise to always be there, to help Thomas and her raise this child. I'm supposed to be the one who ships the little boy toys from far-off places, gives him advice when his parents are difficult, takes him to Saturday movie marathons. I am supposed to be there if something happens.

I want to do this.

But at this moment, this time, for me, with one parent scattered across the ground and another fading, my faith in a positive future feels strained. Nothing seems certain anymore. I look at the skyline and only see the shadows of an EKG line. From high mountaintops, the horizon looks like a frown; a keeper of fortunes and fate; a teller of tales; a gypsy of a changing climate, of a world no longer built on good will.

Every day I see my mother. Often, she is residing just outside my vision, a streak of red that I intuitively recognize as the shirt she wore most often in her last weeks. A plain, red, button-down shirt. She liked blue jeans, a tucked-in red shirt, suspenders.

My mind runs with this blur, this false image hovering in the periphery.

In Skagway, for several years, I was a medic and firefighter and search-and-rescuer. Once, the summer before she died, I assessed a patient who wore suspenders.

It was raining in Skagway, and I was getting off work when my beeper sounded shrilly. The early-evening page was for a medic. I hopped on my bike, weaved down State Street, and flew into the ready room at the fire hall.

Twenty minutes later, I was belted at the waist and chest and packed into a helicopter, flying over town and Dyea, a neighboring valley. The ground was wrapped in mist; the clouds were low; the air was bumpy. Jesse, the pilot, was a man I trusted, knew. I'd worked with him in years past, when he flew me daily out to the glaciers surrounding town. Jesse was a solid pilot who'd fly us through anything.

The call we had received was for a potential poisoning. The patients were part of a film crew that was out on the Chilkoot Trail. The trail runs through the mountains outside Skagway—it starts at sea level in Dyea and twists over thirty-three miles to Bennett, British Columbia. There are nine main camps spread out along the trail, on both sides of the border. Outside of Sheep Camp, the last camp on American soil, it appeared that the patients' stomachs got the better of them, and they started looking for plants growing along the trail to eat. One patient recognized a plant that looked like a lily. Five patients had fallen upon it, shoving it into their mouths. They only stopped when the burning started. It wasn't a lily. Instead, they had ingested one of the most poisonous plants in the area: false hellebore.

False hellebore is baneful; a plant my mother would label "an asshole," a plant that slows the respiratory system, calming the heart into a soothingly slow rhythm leading to death. It looks innocuous, appealing, even tasty. It brings death.

Wendell Berry writes that "it is in the destruction of the world in our own lives that drives us half insane, and more than half." The individual world we recognize, across all planetary systems—it is fading, and we know this. How do we bear it?

There is an ecology of such destruction that I struggle to understand. Ecology, at its most basic, is the illustration of how living things connect to each other and to the environment. Living things include both the more-than-human world and us. How do we connect to one another, to the living world? And once we've connected, how do we destroy that connection?

In the *Journal of Animal Ecology*, researcher Yannis Papastamatiou published an intricate study about sharks.

Tiger sharks are so in tune, so connected to this planet that they are able to store in their brains mental maps of areas spanning over thirty miles. They've adapted, year after ancient year, for centuries of existence. "Our research shows that, at times, tiger sharks and thresher sharks don't swim randomly but swim to specific locations," writes Papastamatiou.

The sharks are able to remember vast areas, research suggests, by using ocean currents, water temperature, or perhaps smell. They might even tap into the planet's magnetic fields.

What would it be like to be so "of" a place that we could remember in detail huge swathes of landscape, wind patterns, temperatures, or smells? How would this affect the way we would interact with our surroundings?

Wait: isn't this how we already are or were?

Sharks are cognizant of their world. They're connected, plugged in, constantly updating their social media. The world is there in the multitudes of acts they do daily to go about the business of living. It is the same for us, only on a different level. We connect to our world in ways we can't always express or feel. We even spend a great deal of our lives doubting or seeking to understand our connection. Often, we feel we've lost it, that we're adrift, that we can't imagine—or that the connection might not have existed in the first place. I don't know what happens to sharks' minds when they can't find a familiar smell, or magnetic field, or when they discover themselves adrift in the middle of the Pacific Ocean, alone. But we know what happens to their shark bodies when something goes wrong, when their mental maps malfunction or fail. They wash up dead on small island beaches. Connection lost, life lost.

It isn't so simple for us. We don't just die when we lose our connection to the surrounding world. We just keep swimming, albeit aimlessly. We feel uncertain.

Survivability, for humans and everything else, is anchored

to planetary stability, safety, and security. The safety of this anchor takes away the uncertainty, like family. Think about the holidays. Think about being outside when it is dark; perhaps you're cold. Think about the moment you walk into your family's home, and it is warm, and there is something cooking in the oven, and the lights are on, and there are people there whom you love. There is safety there.

I keep one of my mother's sweaters in a chest. At times when I am most dark, I open the lid to the chest, root down to the bottom, and pull out her sweater. It is soft and comforting, and I often lay my face on the yielding material. It is the smell, though, that gives me the most security: her smell, my mother's smell. It realigns me. It instantly gives me a feeling of safety.

Bill McKibben writes that "nature has always provided the 'deep, constant rhythms,' even if, in our turbocharged and jet-propelled arrogance, we have come to think that we are independent of the earth's basic pulses." Nature, simply put, provides us with stability, even if we believe ourselves free of all earthly connections. When we're adrift, questioning our connection to the world, it is with nature that we beat back the uncertainty. It is in our landscape that we find familiarity and comfort. Nature belongs in our extended family; the comfort of returning to a familiar place is just as great as that of digging out my mom's old sweater.

We need our connections, in spite of our ability to ignore them. Today, nothing is as it was. So often, every day, I am told again and again what systems are shifting, what consequences are being observed, what future is limited. The impacts of climatic changes are immense. I often feel that we can't count on anything. That the certainty we used to hold in everything has vanished. The stability of Earth's family is breaking down. Our ecology is unraveling, the equilibrium is rocking, and our bodies and minds can't handle this. We cling greedily to the person we know is dying.

McKibben, again: "There is the sadness of losing something we've begun to fight for, and the added sadness, or shame, of realizing how much more we could have done—a sadness that shades into self-loathing."

We are going to lose so much in this world. Much of what we lose we will not know we had until it is gone. This staggers me. I am going to lose my father on the heels of losing my mother. The connections I have forged with my parents my entire life must fade. It is the knowledge of this reality that I see also in the discussions we hold about climate change. We know what is at stake, what we stand to lose, the extensive conversations we have not held yet with our dying parents.

The United Nations released refugee information for 2003 and noted that, for the first time in history, the 25 million environmental refugees exceeded globally the 23 million war and political refugees. Twenty-five million people no longer depend on the stability, safety, and security of the planet. How is this tolerable?

At times, I am able to conceptualize what happened to my parents through the scientific lens of climate change and our planet.

We know that climate change is eroding the foundations of everything human society holds precious on this planet. We know the cause of climate change.

But what does this knowing provide, or do? What is the result? Where does it lead us?

I knew that one illness was paving the way for another within my mother's body. They built upon each other, one and one and one and one. I knew the catalysts, the names. Cancer. COPD. Pneumonia.

And so, greater atmospheric carbon counts feed greater storms and deaths and pollution and the floating trash pit in the middle of the Pacific and people's shifting homes and on and on and on. This is a reductionist's dream, a causality nightmare. Anyone who says cancer is isolated is too short-sighted; anyone who says climate change is a problem of the environment is equally so.

Is the outcome not imaginable?

Yet the end never feels like the end until it has come. Will we know we have gone too far when it is too late? Or will we keep settling?

Where is the end? Lately, I haven't thought there will be one. Rather, that the end is this fictionalized thing we imag-ined so we wouldn't have to deal with the here and now, the reality of this very moment and the climatic changes that are impacting it.

Were the deaths of my parents the end of the story? A story?

Increasingly, the news relates the gossip of the world, and it is looking grim. The phrase "when climate change hap-pens" seems common. As if climate change is a well-dressed monster hanging out just offstage, and at some point in the near future, it will come strolling in and surprise us all.

Often, speaking over the years with my mother's doc-tors, they weighed quality of life versus quantity of life. Is the length of life as important as what we're able to do with that life? Is it better to be hooked up to hospital machines for months if it means we are alive? Or do we go home and enjoy the time left to us, free of restrictive machines? I am uncertain whether ice sheets, or trees, or fringe species, or oceans and evening winds get to make this choice. A global cardiac pump looks to me not unlike an oil well.

Did the brown trout in Norway make this gamble? Shave off parts of their bodies in the hopes of lasting through the winter and risk not having enough strength to get through spring, or go into winter whole and fat and try to subsist?

Is it the same for us? Should we weigh our collective quality of life against its quantity? Science tells us that climate change has an estimated forty-year gap between a cause and its effect. Called "climate lag," there is a profound delay between actually emitting carbon and experiencing the consequences of such emissions. The planetary changes we've registered today, from melting icecaps to mass extinction to increased weather extremes, are potentially products of the carbon emitted in the 1970s.

But then, at the same time, every day we're learning new things about the effects of climate change. And that the lag isn't really forty years. Perhaps it is five, or ten, or varied, or scaled. The effects and experiences of climatic changes are wildly similar across the planet, and, just as often, they are dissimilar.

Perhaps the consequences of emissions today won't be seen until 2050. Should we worry about what 2050 will look like? Or should we focus simply on arriving at 2050 as a species? I will be sixty-eight years old then. Today, that seems unimaginable. My sixty eighth birthday might happen in a world running on clean energy, with climate change slowed down, or, conversely, the world might be a whole hell of a lot worse. Or our world might be somewhere in between. Where it won't be is at the end.

My parents do not end at their death. Climate change is not the end.

I'm reminded of an article about whales I once read in *Earth Island Journal*, about how they are remarkably intelligent creatures, full of instinct and grace and beauty and desire. They require the same air that you and I do. I'm attracted to that shared necessity of living; these deep-ocean-goers and I need and taste and touch the same air.

Again and again, I find parallels in the world we inhabit.

Whales swim right up to the surface of the ocean, breathing at the interface between water and world. While filling their lungs they're at their most vulnerable, dodging boats and logs and sharp rocks and harpoons.

In the spirit of safety, often a whale will propel herself upward with extra fin force and poke her head out of the water. She'll suss her surroundings. She needs to raise her nose just high enough to clear her eyes of the ocean. The technical term for this is "spy-hopping." Sometimes whales spy-hop to make sure areas are safe; other times the whales appear to be inspired by curiosity. Something catches their eyes and they investigate, treading water with noses skyward for stretches of fifteen minutes or more.

Spy-hopping is whales' short-term crystal ball. They can check out the neighborhood and predict danger: a boat, an incoming tsunami. There are times when I wish I could spy-hop to 2050, drop in on my birthday party, see what remains of my future life. I want to see what the world is like, what the payoffs are from today's choices.

What if all of human society spy-hopped forward? What if everyone peered into the future and found a planet smoking and grim? In the fifteen minutes we treaded water and peered into the future, would this force a change of course?

Looking ahead requires courage. Imagination. It is so much easier to keep afloat, to bob along at one's current pace. Sometimes, courage deserts us entirely: I cannot speak reasonably yet to my brother and sister of what it might look like when it is just us, three left from a family of five, too young to navigate without a paddle and a rudder.

I know the hope we portion out is unlikely to lead us to a perfect world. Just as I held out for cures and transplants and more time with my mother, I find myself having the same hopes for my dad. I need more time with my father. Have we considered radiation? Is there a way to remove the lesions surgically? Hope lingers.

I read in science articles that humanity needs more time. More time. We need to find a cure for climate change, to investigate geoengineering transplants, to push governments to back clean energy initiatives and policies. We need more time.

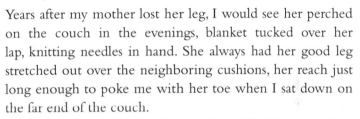

Years after my mother lost her leg, I would see her perched on the couch in the evenings, blanket tucked over her lap, knitting needles in hand. She always had her good leg stretched out over the neighboring cushions, her reach just long enough to poke me with her toe when I sat down on the far end of the couch.

My mother loved to poke me. Especially if I was reading. It made her laugh when I was obviously on the last few pages of a book, sucked in, intent on the completion of the story. I'd read a few lines then get jostled out of the book when her foot bumped my hip. If I ignored it, she'd repeat.

She parked her prosthetic leg on the left side of the couch, within her reach, leaned up against her knitting basket. Sometimes I would see her rub her stump, brow furrowed. Other times, almost reflexively, she'd rub her toes along the empty spot on the couch where her left foot should have been. As if she were scratching the foot that wasn't there.

The closest I ever came to understanding the loss of my mother's leg was when I woke up from throat surgery in a hospital in Pretoria, South Africa. Six weeks previously, I'd begun my service as a Peace Corps Volunteer in Zambia. While settling in to the country I'd picked up a foreign mold that took up residence in my tonsils. The mold aggravated its new home to such a degree that my tonsils swelled and swelled, pushing back into my throat and forward into my mouth. The Peace Corps sent me down to South Africa for medical attention.

I remember feeling a cold hand roughly shaking me, fingers splayed out across my chest. A voice I didn't recognize told me to open my eyes.

As soon as the weak, fluorescent light filtered through my crusty eyelashes, I felt an immense wave of sadness. Loss. My brain wasn't fully awake, but I sensed that something was missing. Slightly acetone-tasting panic settled over me, commanding my heart to beat. I moved my limbs, and everything checked in: alive, sensory. But again, something was not right, and trying to order my brain to focus was like pushing a wheelchair through slushy snow. I couldn't focus.

The nurse leaned over and flashed a bright light into my eyes. I saw white spots. The sound of her pen against her clipboard jarred me.

It took two minutes for me to remember, and my memory was only jogged because in those two minutes the pain grew exponentially until I was certain there was a forest fire burning in my throat and chest.

I knew, before the actual details arranged themselves properly, that something belonging to my body was missing. I zeroed in on the pain in my throat and panicked. What if the doctor had cut my throat away? I flailed. I was saved when my senses kicked in and I could feel the oxygen tube rubbing my neck. My throat hurt because my tonsils had been removed.

We're taught early on that humans have five senses; I remember memorizing them in grade school. We taste, smell, hear, see, and touch. But that's just too simple. We are far too perceptive to be limited to just five senses. Humans also sense balance, pain, time, temperature, direction, death, ill intent, and loss. I know when I'm going to fall over, often eons before my body actually goes. Many times in my life I have felt my skin prickle, and I've turned to see someone staring at me.

I used to wonder whether my mother knew that her leg was gone before she woke up. Did she have to be told, or did she know? But then I finally understood. I felt a kinship with my mother, lying in bed during the days after the surgery. I knew then, at least in part, what she might have gone through. Stretching my mind across geographic and political boundaries and jumping oceans, I reached out to her back home in Washington. I touched her with my mind. I don't know if she ever felt me, but I prayed she did.

The morning after my mother died, I woke up in my familiar childhood bed, warm under layers of covers and blankets. But this time, when I opened my eyes and let the gray light in, the loss I felt was not unnamed. It felt like a different type of loss, the loss of a soul, slamming over me like waves of concrete.

I can't imagine how much my mother suffered from the loss of her leg. I often could see the signs in her of the body-loss pain, but the other stuff—the phantom-limb pain—was a type of loss and pain she kept to herself. She spoke about it rarely with me. I saw evidence of it in her movements, most often when she assumed she was unwatched.

Phantom pain is tricky. The National Limb Loss Information Center estimates that 60 to 80 percent of amputees experience phantom pain, which is loosely defined as perceived pain that is associated with a body part *no longer there*. However, it doesn't just occur following the loss of a limb. It has been reported in patients who lose an organ, or a tooth, even after the removal of a breast. The pain is intermittent, coming and going with no indication or stimulus. Patients often don't report it because they assume it is impossible to feel pain from a body part that is no longer there—and they draw the obvious conclusion that the pain is a trick of the mind. Historically, phantom pain was treated as a figment of the patient's imagination and doctors thought it was caused by hallucinations, or possibly a ghost.

I've never experienced phantom tonsil pain, nor have I been haunted by the specter of my tonsils, dragging chains and moaning at me in the gloaming. But then, I am not sure if I ever actually felt my tonsils. They weren't on the daily landscape of my consciousness. I do feel, daily, phantom family pain. I am surrounded by evidence of my mother, and it is easy, at times, to believe she is still here, to catch sight of her out of my peripheral vision, or to smell her in the wind.

Right now there are red-winged blackbirds falling out of the sky in Arkansas and Oklahoma, thousands and thousands of them, all at once, and no one seems to know how to explain this loss of life. Scientists on the evening news say that it is completely unrelated to the sudden mass death of fish in the same area. Other scientists speculate that maybe all the birds contracted a disease and died. At once. In the sky. And it is not too far of a stretch to look at the sudden hive deaths of entire bee families. Though the cause has long been unknown, lately it appears that insecticides have been killing the bees.

I'm waiting for the ghosts of each of these birds, bees, and fish to begin haunting us, specters in the periphery of our environmental consciousness, cawing about climate change.

Today, I wonder what would happen if I were to throw away my dictionary. If I heaved its massive bulk out the window, its pages suddenly called upon to flutter with winged fervor. Would that stop my mother and me from communicating through a series of right angles? I can't tell my family that my mother and I talk. It doesn't make sense.

I once knew a woman who, every morning, asked the

Bible how her day would be. She would ask the question, then let the Bible fall open to a random page. The first verse her eyes fell upon would be her answer. She'd spend all day interpreting that answer. It consumed her, directed her day.

I don't randomly open my dictionary. Rather, I write and read, and when I need a word, I go mining in its pages. It is too often of late that the word I need is one of her words, one with a gray check or underline or asterisk.

There is meaning here, a Tyche-induced moment I must weigh. Of all the words, this is one of hers. We are so intent on proof. Is possibility enough? The news says that Gliese 581g might not exist. NASA has proven Tyche is a fiction.

Francesco Pepe of the Geneva Observatory in Sauverny, Switzerland, told a recent conference in Torino, Italy, that his team could find no sign of the planet Gliese 581g. They had access to the same data that Dr. Steven Vogt's original study had, but their results differed.

"From these data," Pepe told the conference, "we easily recover the four previously announced planets. However, we do not see any evidence for a fifth planet in an orbit of thirty-seven days."

Dr. Vogt did not attend the conference in Torino, but he sent a rebuttal through the media: "I feel confident that we have accurately and honestly reported our uncertainties and done a thorough and responsible job extracting what information this data set has to offer. In 15 years of exoplanet hunting, with over hundreds of planets detected by our team, we have yet to publish a single false claim, retraction, or erratum."

Oh, possibility. Gliese 581g might be out there, orbiting its sun. Circling. Waiting. It is a place, a nearby place, that people could go. Just to be neighborly. We could make the trip in twenty years if we invented really amazing spaceships, if we navigated accurately in space, if we figured out a way to feed ourselves, if we played a really long game of cards. It is one hundred eighteen trillion miles away from my window, where I stare out at the Libra constellation in the quiet of a dark night. I've stared for hours, but I can't spot it with my eyes. But then, I think I might have been there in my imagination.

I keep looking. But it makes me feel worn to search for such old light, old words in old dictionaries. The light from Gliese takes twenty years to reach us. It would take us twenty years to know if something happened out there. If by chance, the planet suffered a major catastrophe and died. Crumbled. Ceased to exist.

Dr. Vogt says that the Swiss team couldn't find his planet because they weren't using the right instruments. It's all about the tools. The American team in Hawaii uses HIRES (High Resolution Echelle Spectrograph) to scan the heavens. HIRES is the most powerful high-resolution optical spectrograph ever built. Over five hundred exoplanets have been located with it, but they've all been too hot, too cold, too gassy—not like Gliese 581g. Gliese 581g is perfect for us, at least along the middle zone where night meets day, where the cold brushes the heat, where it is temperate and just right.

Possibility. In light of climate change, there are so many things possible. We just have to use the right tools, which naturally include our minds, our imagination, our creativity, our sense of what is possible. Then, and only then, is there a chance that we can bridle our uncertainty, bridle ourselves and our forces and our desires and our reach.

What I know about loss is that it walks hand in hand with uncertainty, and uncertainty is the mother of madness. Go left, right, up, down: the uncertainty of the way forward stalls and hinders movement, choice. Uncertainty feeds on the sheer, inconceivable nature of loss. Again and again, I am told of my loss; I am told of losing something and of the sense of missing I must have. But the reality of loss is like the unimaginability of this very moment's existence. I'm here, I know this is happening, yet everything familiar and known and trusted is gone. There is little to prove that this moment is real anymore.

When we lose, lose on a large scale, our foundations go aquiver, broken, dissolved down into molecular grains.

We cease to trust ourselves.

My mother was sick for a long time before she died. Flitting quick as winter light across my mind would march the idea of her death, the notion of impending loss. But it never seemed certain. I admit, even after her death, I did not believe she was dead. I imagined that she had been washed away, that she had flown, traveled, visited old friends, gone to the city, checked in on Gliese 581g. I stared for a long time at the heart-rate monitor, believing it would jump again, believing the line feeding across the screen, flat and innocuous, would, at any second, quake, shiver, beep.

It didn't. And I walked out of the hospital and into the daylight.

There have been moments of cherished unbelief. Forgetting. Not believing. Once, I woke early at the farm and heard the familiar clock chiming away the hours and the rustle of wind on the roof and I, in the fog of morning thoughts, believed I could go downstairs and find both my mother and father burrowing around in their big bed.

I pulled the blankets back, got out of bed, traversed down the stairs, and entered the kitchen. I coaxed a flame from the gas stove to heat water for coffee. I walked back to the living room and checked the wood fire. I gave it a couple of pieces of kindling and one small log. I walked softly so as not to make the cedar floor creak. I cracked the door to let my dad's dog, Roon Toon Toon, outside. I watched as he walked onto the porch and executed several downward dog stretches before trotting off into the yard, tail pluming.

Then, in the enveloping silence of early morning, I walked on quiet feet down the short hall to my parents' bedroom and nudged open the door. There was a brief moment as I did this, anticipation, and each time I did this, I held my breath. Each morning, the same hope; it arrived unbidden and lingered for less than an instant.

But then I looked in and saw the bed, big and wide, and my dad, small and quiet, tucked alone among the quilts. My cat, awake, snuggled in around his knees, staring at me with lantern eyes.

My dad was struggling for his life, and I had an even harder time believing in the possibility of his death. The uncertainty was excruciating. His cancer was quick and stealthy, appearing in his kidneys and moving into his lungs. The drugs worked for a while, but then they didn't, and I took comfort in the huffs of breath blowing in and out of his lungs, the occasional sonorous snore wrestling up from his blankets.

I left graduate school during my last quarter to be there with him. My sister, Sarah, moved home. She walked from room to room, and the look in her eyes was like the bottom of an oil well. My brother, Grant, worked earlier hours in town, getting up in the dark a.m. hours just so he could drive out to the farm every afternoon and stay all evening. We revolved silently around Dad and each other, and we held our breath. We had friends, and thankfully, they came and went. Dad's friend Al watched Dad's cows, petted Roon Toon Toon twice on each spot. Miki made sure the medications were right, that we were eating and sleeping and talking. Chalon washed Dad's sheets and talked softly to him. People came and went, but this very moment was real and unimaginable. My Aunt Lyn arrived and left and came back.

Dad was there and I stared hard at him from the doorway of the bedroom because I was aware that cancer was a fiery liquid raining down from the skies and could wrap him up and wash him away in mere seconds—it could have happened in those seconds when I looked away to make the coffee, or tend the garden, or feed Roon Toon Toon his dinner.

There are times when I am so uncertain that I am certain I am lost. Everything I believe changes in the course of seconds—unknown landscapes ambush me in the early morning hours when all I expected were sleep and dreams. Yet I awaken in a bed at the farm that is momentarily unrecognizable, three unfamiliar walls and a place where, normally, a window placidly sat, but now, instead, a mountain peeks in, mournfully pointing towards a sky long turned blue. Three blinks of my eyes and things go back to their normal places; the mountain backs out of the window and reassumes its normal place in the landscape.

There are things of which I am certain: my father loved and trusted me. But then, there he is, out of his mind because of cancer, yelling that he believes I am trying to poison him with his prescription medications. I back silently out of his room. My sister goes in, gives him his medication. He takes it. Earlier, my dad thought she was poisoning him. We take turns. Dad is recognizable as Dad, but the medications and the cancer have altered him, changed him, and the future is unimaginable.

My sister walks back out of the room, and our eyes meet. These are the events that break families. We were five people, but Mom is gone, and our dad is sliding away. There will be only three of us. We must, at some point, talk about this, re-imagine a future family of three.

She walks out on the porch. I go into the kitchen and ruffle through the cupboards. Since neither of our parents drank, there is usually no alcohol in our house. I dig deeper, find a bottle of wine. I grab coffee mugs, the wine bottle, and I go out onto the porch and sit down next to my sister. She doesn't drink. But I open the wine and I pour it into the mugs and I hand one to her and she drinks with me. And we sit there, and the summer's evening light shifts down, the frogs begin to talk among themselves, and the wind creaks through the windmill. She puts her arm around me, and Roon Toon Toon appears to rest near us. Sisters, we watch the evening and drink our wine.

I remember that girl I'd loaded into the helicopter, one of the five actors who'd eaten false hellebore in Skagway. She was scared, terrified. She wanted her mother. We'd put all five of the patients into the helicopter to fly them to the clinic. She was the worst off. In the air, I struggled to monitor five patients as the helicopter lurched and tilted in the rainy clouds. The first time I tried to take her vital signs, I couldn't detect a pulse, even when I shoved two fingers against her carotid artery. I thought I was doing something wrong. I tried again, but I still couldn't find a pulse. She was awake, aware. I moved on to the other patients.

I have held the root of false hellebore in my hands. Hallucinations accompany ingestion.

After getting vitals on the other four patients, I came back to her. I felt again for a pulse. Nothing. I watched her, asked her what she saw. Her pupils thinned. Her respiration slowed—dangerous.

I tried to hold her hand, to comfort her, to keep her warm and focused. We landed at the TEMSCO base and people surged towards us from every direction. We loaded all five into the ambulances, drove the length of town, and admitted them at the clinic. Skagway's clinic is staffed by the best people: cool and calm, deeply caring.

When we are at our worst, we need mothers. Every time I returned to the patient's temporary room in the clinic, her eyes lit up; she was glad to see me, to recognize a face. She regained color, started to feel better. She kept asking how I was doing, what I was doing, where I lived; she wanted to build a bridge with me, a connection.

I have never taken false hellebore. I do not think I need to imagine hallucinations. Already, there is too much in my life.

I said the wrong things to Jordan, my friend who had asked me to be her baby Cooper's godmother. I should not have paused. I should have agreed immediately. I should have said this on the phone:

"Jordan, you are going to be a good, strong mother. Your baby will have a godmother in me. I will do this. And it will be okay. I will walk the beach with you and baby Cooper; I will teach this child about ice and forests and hope. The world this child inherits will be beautiful. It will utilize green energy, it will be a fount of possibility and opportunity and be just and equal and hopeful. There will still be oceans, clean and blue, forests with trees older than grandparents. The air Cooper will breathe will be cool and clean. This baby will gaze into the night sky and discern mystery. Nothing will happen to you, or your baby. I will be here: I will imagine and make this future for your baby."

I didn't say that. I couldn't. I could see my friend on the other end of the phone, one hand permanently ruffling and unruffling her brown hair, the other holding the phone to her ear. I knew she was wearing a silver ring. Likely, she was pacing.

But in the same breath, the one with which I was going to say "no," what came out instead was "yes."

I think that when we look around and cannot find any certainty in this world, we must find it, imagine it, in one another.

I GET PANIC ATTACKS. WHEN MY MIND HAS HAD ENOUGH,
my body folds. By "fold," I mean that I tend to bend like a
lawn chair at the waist and lock. My brain insulates itself by
building minor blockades; ordinary messages, like "swallow,"
or "move left hand," or "breathe," get sidelined, bounced off
the blockade and returned to sender.

In the days of waiting, as my father grows worse and
worse, as he's retreated from the porches to his room, as walking is too hard and he is too tired, as his body wastes away
while his mind stays alert, I panic.

Generally, my knees draw up tightly against my chest.
There, they press down on the complaining lungs, tightening ratchet-style until the veins can't bend anymore and the
blood flow to my toes is compromised. My brain is aware
that this is a bad thing, a bad move, a bad position, but it
does nothing to help. My brain chooses not to answer any
requests. Breaths arrive in short bursts, beaded thumbnails
tacked along a narrow string of saliva.

I can't predict when I might fold over like a lawn chair. I
carry the effect, though, for days afterwards. My heart thun-
ders at the thought of it happening again. My hands feel
arthritic, aged. Often, in the midst of a fold-over, my fingers crimp, collapse—the unbending, releasing cellophane

sounds backdropped behind the machete-chopped thuds of my joints popping.

Once, I had two fold-overs in two days, a personal record. The first time, I was yelling at a friend in the swampy a.m. hours, too exhausted and heartsick to choose words with care. He was standing in the middle of my apartment home in Missoula, fiddling with pens in his pocket, not making eye contact. He was leaving and I was screaming at him to stay, and we both knew it was done. In the middle of a particularly vilifying statement, he stormed out and my back seized, the muscles along my hips went taut, then tight, and I became immobilized.

For the second fold-over, a day later, I was opening a jar of rice in my kitchen and thinking about driving home to the farm. I was warned by a twinge, a revolt of the muscles in my calves, warning enough to drop the glass jar to the top of the table before sinking down, melting into a pile. I lay there, staring out the window at the hill laced with snow and winter wilt across town, scared and tired, but sensible enough to wait this one out.

To get my body to relax, I blanked my mind. I blanked it and thought about just one color, red, and I held that color and I held it and I waited.

It took me a while to figure out what had caused that second fold-over. I was multitasking before I went into the kitchen to open the rice, simultaneously ordering flower seeds online for my dad, reading the BBC reports about the Arab Spring in Tunisia and Egypt, and reviewing my climate change seminar notes. I knew before I stood up from my desk that I was agitated, bothered. There was a tightness in my neck, a metal taste in my mouth. On the wall by the kitchen door was a picture of my parents when they were younger, and I would look at it each time I went through the doorway. This helped me, most of the time, combat the image that would spring unbidden into my mind of how my mother looked after she'd died. Sometimes, my mind will switch parents, and I imagine how my father will look.

I glanced at the photograph after I got up from my desk, and then I walked into the kitchen and picked up the Mason jar, but instead of seeing the rice I simply saw death, and it was too much. I folded over.

About a year after my mother died, I heard two prominent environmental experts speak in Missoula about climate change. Both outlined the problems, issues, and facts. They were knowledgeable, informed, serious. In their presentations, climate change looked grim, insurmountable. At the end of the lectures, each expert offered solutions. One said, "joy." The other, "passion." Joy and passion will answer climate change.

When pressed, one expert offered switching to low-energy light bulbs as an additional solution.

Frustrating. I feel that most climate change–related lectures, or books, or podcasts I've been listening to lately fully elucidate the problems we face in excruciating detail—but they miss the focus on the after part, the "what's next," the future plan, the catalyst for action. With all due respect to these experts, so much more is needed. What the hell happens after the problems have been identified? Where do we go from here?

In the face of the enormity of climatic changes sweeping across this planet and the equally enormous range of discussions about those changes, from everyday, lived climate change experiences to the worst-case apocalyptic scenarios we've heard over and over, why, in the face of all this, are we being told to change our light bulbs? Or to follow our joys while pursuing household planetary interventions, such as checking our tires, adjusting the thermostat, or unplugging the toaster when we're not home?

It feels banal.

It is like the time when my dad's doctor took a half an hour explaining why he should keep an eye on his hangnail. To the man dying of cancer.

Climate change is massive, it is real and here, and it manipulates everything. As an active member of human society, it is belittling to be told that my capabilities extend to, and end at, a light switch. Or worse yet, it is well known that our individual lifestyles, from driving our inefficient cars, to supporting politicians funded by oil corporations, to demanding the market provide more and new and even more new to buy, directly contribute to and exacerbate climate change—but I am advised by the experts to find and follow my own joy or passion. This does not work. Not only does it provide a narrow glimpse of climate change, but it also leaves all those who went out on a Friday night to hear about climate change feeling disempowered. Frankly, these beacon scholars of climate change should be using their scientific soapboxes to encourage us to get off our bottoms. The last thing our society needs as a take-home message about climate change is one extolling the virtues of what amounts to more myopia.

This is not the time for me to focus only on following my passion, on discovering my inner joy. My inner joy leans heavily towards glaciers, and I don't think such joy affects glaciers or the climate, one way or another. This is not about you or me. For once, individually, we're not the focal point. This is about our home and how we, as a group, act in that home. This is about how we dare to imagine a different path forward, a different future. This is about our children treading after us, the landscape we walk on, about the brilliant blue, blue glaciers slipping off into the unknown.

I like to call my panic attacks "fold-overs" because they scare the shit out of me, and adding a little humor to the problem helps me cope.

I've been thinking about panic a lot lately, since it seems my fold-overs are increasing in regularity. We panic, the self-help books tell me, as a reaction to a sudden, overwhelming fear. Perhaps. But I think it's more complex. Panic, to me, is my body's way of saying "time out." The brain has maxed its available space. There is simply nothing more that can be done, so the body is cinched into stasis, everything slows down, and while I'm on the floor, bursting short breaths and resembling crushed furniture, my brain reboots.

I think a reboot is good. After the last two fold-overs, I spent a great deal of time mentally sneaking around what I think caused them. I'm thinking, analyzing, imagining my landscape. It is a new start.

Sometimes, panic is an appropriate response. It gets you out of danger.

I wonder what would happen if the whole human race panicked. Together. Right now. If, in the US, people saw right through ExxonMobil's platitudinous propaganda campaigns, or saw firsthand the volume of money passing between the right-wing industrial conglomerates of the Koch brothers and various elected officials and political departments. If the science and facts of climate change were accessible and digestible to everyone, and then everyone made meaning of that information and shared it with their families, their neighbors, their networks. What would happen if every human being on this planet could look at one another and say that climate change is happening, that it is largely caused by human society's increased greenhouse gas emissions, and that it is solvable by reducing said emissions and utilizing clean energy? All of this is entirely within our capacity and grasp. But if we don't move now, and I mean now, this planet will be unrecognizable to our children. Our children, and their children, and all the generations we hope will come after us. What if we could say this, each of us, and as we did, what if panic bloomed in our stomachs and hearts?

Would this prompt immediate action, or would we give in and fold over?

I've noticed that, in the midst of my own fold-over, my panic attack, my mind resigns quickly. I'll try to move one of my hands, which has morphed into a claw, or stop my knee from wedging itself under the ribcage. These efforts bear no reward. I'll struggle for a second or two, and then my mind will pipe in and say that there is nothing I can do. *You are, actually, stuck here for the rest of your life. You are powerless to change, to free yourself.* Sometimes, I want to tell my mind to shut it: "Brain, you are not helping."

In an interview with the *Guardian* in late March 2010, James Lovelock, English scientist and loud, respected environmental voice, went on record condemning the human race: "I don't think we're yet evolved to the point," Lovelock said, "where we're clever enough to handle as complex a situation as climate change."

Thanks. I'll continue to be immobilized.

This is not helpful. In the face of a problem we all share, where the result of failure is losing everything, saying we are simply not intelligent enough is a complete waste of words. It is so easy to say that we're not smart. How can Lovelock, so brilliant, display such a tremendous, terrifying lack of human imagination? I understand that he is disillusioned with the way human society has dealt with climate change so far. I am, too. But I am still here. And so are you. And we're all capable of imagining and making a difference if we choose.

Lovelock sounds like he has given in. But what should

we do if we're really not evolved enough to face climate change? Sit down, play cards, and wait, generation after generation, until humans are adequately evolved to be able to juggle enormous planetary problems? Should we get up each morning, scratch our heads, and say, "Nope, can't handle that today"?

It is so easy to give up. To say to yourself, "We are not smart. We cannot act together. It is beyond us. We will simply remain here, on the floor, with our knees tucked into our ribs."

When my mother died, I had a choice. Let my grief drown me, or keep going. Stay in bed, or get out of bed. My friend Mario was right. Think of how much I would miss.

To this day, I am not sure whether I am evolved enough to understand adequately the enormity of my mother's death, or of climate change. But I am trying, every second. I am imagining, and I am walking into a future ever so dark and uncertain and unknown, with my hands reaching out for meaning, for understanding, for a way forward every day.

A while ago, after my mom died, I walked into our pastures at the farm with my dad to check on a cow. She had isolated herself from the herd and was tucked down in a grassy bowl. It was calving season at the farm, so we walked out to see whether we had a new calf.

Walking the pastures with my dad is akin to strolling into a sanctuary. I've been roaming these spaces with him my entire life, and as we walk, elastic bands of experiences and memories and friends and family rebound and rattle and ricochet behind us. I find spiritual peace in the damp turf, in the foggy cedar bogs that line the fences, in the company of my father. My dad speaks in low tones of this year's rotation strategy, which pastures the cows appear to favor this season, where he saw an elk the evening before. Our feet sink down into the mud by the gate, and he tells me about the new drainage system he'd like to put in place, one with a bobble sump pump up by the main barn. In two, maybe three years, he thinks, he could have most of the runoff redirected so it won't gather in this low spot.

My dad has three to six months of life left. He knows this, and I know this. He isn't in denial, though; he appears to have found peace with inescapable reality. And so we walk out into the pasture. There are coyote tracks and damp spiderwebs stretched out on the fence lines. Dad talks about the cows, about the registered brand he is filing with the state. We approach the downed cow, and she lurches to her feet, revealing the little, black calf beside her. The placenta hasn't passed yet, and the calf's hooves are still white. It is minutes old.

We check it: my dad's hands are calm, soothing. She's a little heifer, still steaming.

What I know is this.

If I focus on the pending death of my father, I miss the life he has now and the life he has lived and the moments, hours, days we share like this one. If I focus on his impending death, it defines him, his life. My father is much more than his death. Death is one of the details, but it does not define everything about my dad or my mother.

There is so much more than the details directly within eyesight. I think, at times, that we can focus so much on this enormous, dreaded, planetary phenomenon, climate change, that we overlook our capacity to engage realistically with it today.

My father, I think, could turn inward, could become so burdened by the knowledge of his death that he stops living. Every loss requires choice. He could turn away from us, from his family, his farm, his cows; he could give up.

I have friends back in Alaska who live off the grid, harvest most of their food from the landscape, and live quiet lives out of the way of the bustle of humanity. When I visit them, I feel relaxed and peaceful and slightly puzzled. They have turned their backs on the world deliberately. They don't want to participate in a society of which they don't approve. My friends just want to be left alone. But there is a luxury to unplugging from society, and an accompanying shortsightedness. On this planet, this place that we share, I don't believe we can hold on to those individual luxuries anymore. It simply is not enough to pursue individual "joy." We've moved beyond that.

If we turn away from each other in the face of climate change, from our communities and our nations and our world, we will lose it all by individual acts of solitude and survival. Climatic changes are these enormous problems that we understand to be lumped together as "climate change," this mother of all problems, and it is all-encompassing.

If we focus just on energy-efficient light bulbs; or on individually moving off-grid, back into the woods; or on a pending death, we're never going to be able to address fully the challenge before us. We'll get bogged down.

Focusing on joy is beautiful. But it isn't enough.

Climate change will not care if I lived a carbon neutral lifestyle, or if you did. Or if the entire city of Missoula or Seattle did. Or if just America did. But if the whole planet did, that would be another matter. The scale of climate change is planetary, so the human response to it needs to be planetary. And that human response needs to consist of coordinated human acts, on multiple scales, ranging from individual to local to regional to national; of multiple governments acting together, in concert, as a planet. Individual actions in the face of climate change are only meaningful when they occur in concert with the actions of the rest of humanity.

For so long, this country has defined itself as the home for dreamers. Come, stand on this blessed soil, and pursue your dream. Yours. We've wrapped ourselves around the holy idea of the sacred individual. But that rampant individualism has led us down into a viper pit with climate change. Today, we must think harder, imagine differently. We have to believe anew in the sacredness of the collective dreamer, the unlimited imagination of a nation and a planet.

We know what we have to do. We know that climate change is going to disembowel our species, that reducing greenhouse gas emissions and utilizing clean energy is our most hopeful strategy, and that we have to act now.

But we haven't yet. We haven't because we're still stuck and because we're afraid of imagining our country, our world, our species, differently from how it is today. It is so easy to believe the worst in ourselves. It is so easy to say that we're not smart enough, that this country is going to the dogs, that people are unchangeable, that the planet is fucked. It's so easy to say that we can never grow beyond war, that social inequity will always be the norm, that the altar of wealth will forever be the only one worth bowing to. But what is easy is not always right. And as a people, in order to do anything meaningful, we are going to have to think better of ourselves. We have to look beyond my dad's death and see his whole life.

There is power in owning up to our having a hand in the making and building and exacerbation of climate change. Recognizing, however, does not only imply blame. It says, instead, that we did this. And we don't have to anymore. We can move on. We forgive ourselves. Then we demand better of ourselves. We re-imagine our future, one raised so lofty into the skies of idealism that our eyes grow sore with strain and wonder at its possibilities.

Climate change is not going to go away. But it is different from panic attacks and the loss of a father. I can't do anything about those two, but there is plenty I can do about climate change.

Auden Schendler, sustainability director for Aspen Skiing Company, writes that "climate change offers us something immensely valuable and difficult to find in the modern world: the opportunity to participate in a movement that, in its vastness of scope, can fulfill the universal need for a sense of meaning in our lives. A climate solution—a world running efficiently on abundant clean energy—by necessity goes a long way towards solving many, if not most, other problems too: poverty, hunger, disease, food and water supply, equity, solid waste, and on and on."

"Climate change doesn't have to scare us," Schendler continues. "It can inspire us: it is a singular opportunity to remake society in the image of our greatest dreams."

Schendler inspires me.

I am not immobilized by fear. I am inspired and empowered by this challenge. There is so much we have the capacity to do that goes beyond using energy-efficient light bulbs. Please understand: I think changing light bulbs, checking tires, and adjusting thermostats are all good actions. But they are not enough. They are not solutions.

Back to those experts on climate change: You have a willing audience. Do more. Explain the problems of climate change, and by all means, extol the powers of joy and passion. But please, I beg, don't stop there. Give us answers. Demand that we imagine solutions. Tell us to join together. An age-old idea. Before climate change, a single person may not stand. But a planet of people, standing together, passionate about the change they envision: this is powerful.

Governments will put the onus on us to reduce our in-dividual energy use, but they won't make meaningful policies and plans unless they are pushed to by their constituents: us. Bill McKibben famously says, "By all means, go screw in that efficient light bulb, but then, go screw in a new senator."

The biggest assets today in the fight against climate change are our feet and our mouths and our minds, our wallets and voting choices and buying decisions, our words and our imaginations and our physical acts. Somehow, in this country, there is a belief that gathering together for a shared cause does not work. The momentum created by the Civil Rights Movement or the Vietnam War protests or the Montreal Protocol is seen as history, unrepeatable. A day off work to march or boycott risks us our jobs.

Because we do so little together each day in the face of climate change, it can often feel as if nothing is being done, or worse, that nothing can be done. As though there is no way through. But the perfect answer to climate change remains an immobilizing myth that sits on the mantle and stares at us, all marooned in armchairs. The imperfect, par-tial, wild and radical solutions to climate change form a web of approaches enacted each day by a host of people (including you and me) with feet and mouths and minds. It is too easy to focus on the negative, to demand perfection before the preliminary trials, to exist perpetually dissatisfied when successes come in increments, drops in the bucket in the face of a tidal wave. Rebecca Solnit reminds us that "perfection is a stick with which to beat the possible."

We must gather together, physically, mentally, in the physical world and in cyberspace. A part of each day must be devoted to imagining a better future. Those with the loudest voices, scientists and writers, philosophers and artists, celeb-rities and politicians, speak up. Climate change experts, keep doing what you're doing, researching, developing, talking about this planet, our environment, climate change. Keep the conversation going. But while you're up there at the pulpit, while you're talking about ice sheets and floods and radiative forcing and carbon caps and joys and passions, start challenging us to imagine harder; to get organized and creative and loud; to use Facebook, YouTube, Twitter, Google; our mouths and minds and feet to build towards a multitude of imperfect solutions that all add to the greater whole, the better future, the world with reduced greenhouse gas emissions and much cleaner air.

Tell us that this is about our very dignity as a people. Dare us to walk out from our jobs, to arrange boycotts, to recall from office obstructionist public officials and naysayers who cloud the truth and bar the way. There is no debating the truth of climate change. It is here, and we have the capacity to respond to, and grow from, it. Tell us that what we do as a nation matters, that we are capable, that our only limitation is the breadth of our imagination.

My dad requested flowers for spring: for what may be his last spring, many flowers in the gardens around the house, in the side gardens beside the apple trees, in the front garden with the vegetables, in the pots and boots and hanging baskets.

When we walked back up the pasture from the newborn calf, he talked about the flowers he loved. We stepped through the corral gate with the squeaky latch and waited for Roon Toon Toon to catch up with us. We checked the water troughs and talked about the gutter that needs fixing on the main barn. My dad pointed to a place near the windmill where he wanted to put in some small, pretty trees in the next couple of years. We talked about the daffodils that would look great there.

The Talmud tells us, "You are not expected to complete the task. Neither are you allowed to put it down."

And so this is what I do.

In Missoula, in my tiny apartment, before spring comes, I order online pounds and pounds of sunflowers, Indian blankets and teddy bears, black oils and Mexican reds. I order bachelor's buttons, chrysanthemums, dahlias, alyssum, marigolds: flowers and flowers and flowers.

When they arrive, nestled tiny in paper packets, I drive the long drive home to the farm in Washington, and I plant them in beds of sandy soil and nurture them to adulthood. I do this because it is doubtful my dad will see this summer, and if he does, what he must see is color and flowers. Before he meets his maker, he must be surrounded by the flowers of his life. And so, in Missoula, I order seeds online and go to the kitchen to make rice, but instead, I fold over under the weight of this, of what has happened, and what the future holds. And when the panic attack is through, and I stand up, I walk back into the living room of my apartment, sit down at my desk, and order more seeds. Cornflowers and calendulas and larkspur. I do this because, for me, there is no other option. I do this for my dad. For spring.

There is a beach south of Skagway, Alaska, at Glacier Point, that specializes in what are called "moon rocks" locally. The stones are of the quartz family, translucent, perfectly white, and ground smooth by the ocean. They look like little bits of the moon, fallen to the ground.

A couple of years before my mom died, I decided to ferry out to Glacier Point. I was restless and in need of new scenery to set eyes on. I hadn't left Skagway in months, which, given the nature of the town, is a long time. Skagway is tiny, about five blocks wide and twenty-three blocks long, wedged into the bottom of a narrow, glacially carved valley. The town is historic, dating back to the Klondike Gold Rush, with a large portion of the present-day microscopic downtown part of the Klondike Gold Rush Historical National Park. The population fluctuates depending on the time of year, and in winter, there are around five hundred people living on every inch of available flat land in the valley. In summer, that number triples, mainly in response to the million-plus tourists who come to the town each year between May and September. Tourists get off the cruise ships and jam-pack the five streets lined with historical wooden buildings. Many of my friends dress in period-appropriate costumes from 1898 and give tours, wait tables, or drive buses and streetcars. It is often comical to go into our local coffee shop, Glacial Smoothies, and see, for instance, my friend Nicki Bunting dressed in long floral skirts, a corset, a parasol, gloves, and low-heeled, sensible black boots. She always tucks her enormous head of brown dreadlocks, not period-appropriate, demurely into a scarf-and-hat combination.

Needless to say, it can be rather crowded in Skagway at any time of the summer. Hence, I told friends that I was going to get out of town, and headed out for a much-needed respite at Glacier Point. I was reminded to pick up moon rocks. A woman in town used to trade coffee for handfuls of them. She was lining a path on her property and loved the stones' luminescence.

At Glacier Point, I kept an eye out, and as I roamed the rocky beach, staring at the Chilkats, I filled my pockets with the white stones. Free coffee.

But it was near the end of summer, and I never crossed paths again with the rock-collecting woman. Instead, when I left Skagway and returned to the farm in Washington, the moon rocks traveled with me. I gave my mother the gift of small, white stones that glowed in the ghostly light of the moon. She loved them.

My mother placed the rocks together, one glowing heap, in the flower bed along the path to the house. Early every spring, she'd leave off her prosthesis, and she'd scoot up and down the path on her bottom, awakening the flower bed from winter and weeding. She'd clear away the debris drooping over the stones. Their glow never faded, even after years had passed.

It became a habit of mine, in the evenings, when I walked up from the barn to the house, to check the rocks—to see whether they were still infused with lunar light.

The Kaonde people, like so many other cultures, exercise a variety of unique funereal customs when burying their dead, many of which I came to know when I lived in Zambia during my Peace Corps service. Before burial, sometimes, for reasons still not entirely clear to me, men and women would carry a coffin through the village, yelling, shouting, spirit-begging. Villagers would gather and watch, and soon the sound of keening, distressed screams could be heard all throughout the forest. It is unforgettable, a church bell tolling across a silent landscape. The echoes still the heart, groove the brain, burn the eyes. The first time I experienced a Kaonde funeral like this, I felt as if my soul had escaped through my teeth. I had to put myself back together emotionally afterwards.

What I learned from my Kaonde friends is that the coffin of a dead person can magically gravitate towards the person responsible for the death. If this happens, and it has, the villagers will exact justice. Death must be paid for, avenged. Villagers trust the dead. The dead do not lie.

The first Kaonde funeral I saw, I was visiting another volunteer, Doug Evans, who was already halfway through his service. I had been in Zambia for a couple of weeks, and this was my first journey north. I spent my first month outside the capital city, Lusaka, learning the local language, acquiring the skills required for my new volunteer job, and forging bonds with my fellow volunteers. Once each volunteer was assigned to a region, a visit followed. I traveled north from Lusaka through the bright-brown countryside, tasting the sounds of the gentle names of the towns we went through— Kabwe, Ndola, Kitwe, Chingola and Chililabobwe, Solwezi. My assigned area was near the Kansanshi Mine, one of the largest copper mines in Africa, in the heavily forested area just northwest of Solwezi town.

Doug lived about a day's journey west of Solwezi. A day into my visit to his assigned area, we trekked into his *boma*, his town's center, to resupply. It was afternoon, quiet, and the heat was vigilant and watching. We killed chickens and drank Fanta Orange and smiled at the clumps of ever-present children creeping around us. I was so new to Zambia and found it all new and beautiful, and the air was dusty, and the sun was slung low in the sky, pouring horizontal light over the acacia.

When the sounds of wailing started, I remember looking around, curious. I remained at rest in my plastic lawn chair, sipping my Fanta. The wailing of the women did not convey the horror of a sudden, terrible event; rather, it sounded like an unfurling lament. It grew louder and louder. Doug had been warned that there would be a funeral; we, in turn, lingered longer in the *boma* so that he could pay his respects.

The coffin came first. It was carried by many Kaonde people, their fingers and palms and shoulders supporting, moving, floating it along. The women in bright *chitenges*— traditional, brightly colored and decorated fabrics—came after, hazing in and out of my vision in the dust kicked up by hundreds of feet. The women wailed, tore at their breasts and hair, rent their garments. The human cries bred with the surrounding biophony: together, the volume and urgency increased in waves.

Later, we went to the area of the *boma* where the funeral-goers had gathered and suffered together. The coffin rested, for the moment, outside a hut. Inside, the continual sounds of wailing. I sat, then, in the outer rings of villagers crouched in the swept dust. Kaonde men sat on log stools or stood on their feet. The women, singly or in groups, approached the grass hut, wailing, apocalyptic in their grief, their release. Doug went in to pay his respects.

We watched until our bones shook. It was a human tsunami. I fought my own urge to wade into the frenzy.

Afterward, we walked back outside the *boma* to Doug's mud-bricked hut. We made dinner as villagers walked the path out front, heading towards the grief. Evening settled down around us, the air thick and dark. Conversation was strangled. The flames in the kerosene lamps flickered.

In a *New York Times* article, authors Joyce Carol Oates and Meghan O'Rourke discuss notions of grief and mourning. Oates lost her husband, and O'Rourke, her mother.

"There is a strange sort of expectation," Oates writes, "that grief should conform to a general pattern or principle. There are even scientific polls of measurement—what is 'normal'—what is 'extreme' grief. As if individuals are not radically different, and as if even the course of a common disease, like cancer, will not manifest itself differently in different individuals."

O'Rourke writes, "I was thinking partly of laments and widows tearing their hair and rending their garments—things that help express grief's physical intensity—without the mourner having to be embarrassed. I don't know about you, but I often felt embarrassed in those first months of grief. I worried I would cry whenever a stranger was rude to me on the subway; I was angry that on top of my loss I had to be concerned about what to say when people asked 'How are you?'"

Upon the death of my mom, I no longer knew how to interact with people. I assumed everyone, even the checkout lady, knew of my loss. I became awkward and paranoid, framing each conversation through the lens of parental death. I felt people watching me, waiting for me to fall to my knees, to buckle under the weight of death.

I did not.

As Oates writes, there are social perceptions of what it means to grieve. Well-meaning friends told me often that my irrationality was "normal," and that it meant that I *was* grieving. Others, possibly not so well-meaning, questioned my lack of outward sadness. I ignored them and went about my daily routine of classes, phoning my father, and lying on the floor of my apartment telling myself stories.

I became embarrassed about my sadness, my loss. It seemed so many friends wanted to talk about it to me, at me, around me. At times, I became guarded, reclusive. I simply did not want to talk about it anymore. For me, it was enough that I was still standing. Anything more, even mental dialogue, was effort above and far beyond. Once, a month after my mom died, my friend Kjerstin Gurda stopped by and stood in my kitchen in Missoula. She wore blue jeans with a rip in the right hemline and a red, knit, long-sleeved shirt and a dark-blue down jacket. Her blond hair was pulled back in a ponytail, and she had a thick smattering of freckles across the bridge of her nose. I remember all these details because she knocked on my door, came into my kitchen, we said a couple of words, and then, for twenty-five more minutes, we stood there in complete silence and I didn't know what to say. I stared at her, and she at me, and I was so grateful to not have to say anything. She hugged me tightly when she left.

I wonder why we no longer tear our hair in sadness, or rend our clothes, or throw our bodies upon coffins. There seems such release in this act, so much pouring out of not only grief, but also everything else that has been pent up and festering for years.

Crying actual tears is a release, but it is quick and short-lived. I would go for weeks after my mother's death without crying, without a single tremor of my jaw. But then, the waters would rise and the hurricane would hit land and I would drown. And then, nothing. Until another storm arrived.

There is a man wandering around the Andes and painting the mountains white.

Eduardo Gold is a Peruvian inventor and outdoorsman who has witnessed the drying up and disappearance of many high-mountain glaciers. Over the last few years, he has come up with a solution, a way to stop climate change.

Mixing together lime, egg whites, and water, Gold has created a nature-friendly paint, and he's slowly been slopping this mixture onto the sides of mountains in square-acre increments.

"I am hopeful that we can regrow a glacier here because we would be recreating all the climatic conditions necessary," Gold told the Sierra Club. "I'd rather try and fail to find a solution than start working out how we are going to survive without the glaciers."

Gold's thinking is interesting: by painting the mountains white, he hopes to reflect the sun's rays back into the atmosphere and, subsequently, to keep the mountains cool. With any luck, then, the cooler mountains won't cause as much snow to melt, which would result—many years down the line—in potentially enough snow stacking up and compressing that a glacier could be born.

In 2010, Gold and Licapa village volunteers spent two weeks covering four acres of Chalon Sombrero, a 15,000-foot-high mountain. The intent is to re-create snow's whiteness on the rocky slopes. The white paint reflects solar radiation back out into space, rather than allowing the surrounding landscape to absorb the heat and to warm at rates unnatural for the area. Ideally, this will help keep the slopes cool, slowing the rate of annual snowmelt. Possibly, in the future, this could midwife a new glacier.

Gold has received a great deal of global attention for his efforts. He was among the World Bank's winners of the "100 Ideas of Save the Planet" competition, earning $200,000 to fund his project. But others, including Antonio Brack, Peru's environmental minister, dismiss Gold, saying his project is "nonsense" and that "there are much more interesting projects" to fund instead of mountain painting.

The success of Gold's painting remains to be seen, as it will take years to see whether a glacier can be birthed again in the bare mountain peaks. Many scientists are waiting in the wings. "If it's at all feasible," Andy Ridgwell, from the University of Bristol's School of Geographical Sciences, told CNN, "then it is an interesting thing to try. If you can get some glacier mass re-established then you can have that water supply buffering that glaciers provide."

While Ridgwell approaches the glaciers' value from the perspective of their service to the ecosystem, and Gold, from that of their greater cultural significance, I wonder about the act itself.

It's proven, this technique. Paint white a roof, a road, a field, and it reflects solar radiation. This is, theoretically, good Reflecting radiation away from the planet keeps Earth cool, which is the name of the game in engaging and moving forward to combat climate change.

Eduardo Gold knows this.

But it's a stopgap measure, a Band-Aid on the finger scratch while the femoral artery is gushing blood.

In two weeks, Gold and his volunteers covered four acres. Of the Andes. A place of millions upon millions of acres.

I do not believe Eduardo Gold's words. I think he is

aware that his project is not going to stop climate change, that its effects will be minor at best, and that the findings of scientific studies will be inconclusive.

Gold grew up in the Andes, in mountains he loved and learned. He's in his fifties. Over the course of his lifetime, he's witnessed massive mountain glaciers dwindle, deflate, recede, drop by drop. That does something to a man, to a society, watching life disappear. There is powerlessness inherent in bearing witness.

Eduardo Gold is grieving. His landscape, his mountains. But pulling his hair, rending his clothes, beating his breast bloody: what would this do? Would it bring back his glaciers?

Time and again, after my own mother's death, I was told to "act my grief," to love it in good works, to hike it, climb it, meditate on it, scream, cry—to emit it. I was told to grieve in color because I'm a painter.

But when my mother died, I could not paint. I tried to heed my friends' advice. Painting is an excuse for me to surround myself with color. I find that most colors, regardless of how brilliant they are, stand best and brightest when matched with white. White creates reflection. And so I made canvases, small ones and large ones and narrow ones and rectangular ones. I stretched the raw canvas over salvaged-wood frames, stapling the fabric along the back. I spread brown tarps across the carpet and gooped thick gesso, bright, luminescent white, all over the canvases. I spread it with my hands, dying my fingers, palms, wrists, bracelets white-white.

In a few days the canvases were dry, white, waiting to be painted. I couldn't. I leaned them up against the walls of my home. They reflected.

I left them there, day after day. I did not paint them. White is the ultimate color, the final expression, a hue's rending of its breast. It is both a boundary and a limitless manifestation. White bears witness.

How can this not be what Eduardo Gold is doing? He

shrugs into stained jumpsuits, hauls jug after jug of his paint mixture up steep mountain terrain, and every day, he splashes the rocks white. This act is more symbolic than fruitful. Four acres in a mountain range, quick to wash away with the rains—this is his witness.

Gold cannot force the Peruvian government to endorse clean energy or reduce greenhouse gas emissions, nor can he make the world do so. He cannot stop climate change, nor could he stop the glaciers that bordered his home from shrinking. What he can do, however, is engage, begin, walk, paint. His other option is to go home, sit alone, and mourn the ice in the solitude of his house.

I remember the second and last Kaonde funeral I saw. I was visiting a Kaonde woman who was my friend, Beatrice, and her family. I sat outside her mud-brick home, shelling beans and laughing as her seven children screamed Kaonde and English words at me. I'd respond, sometimes messing up on purpose, other times not, and they'd wriggle in amusement and throw their bodies in complicated loops and rolls and jumps.

Beatrice was one of the few women in the village who wasn't afraid to touch my blond hair, to teach me to sit correctly on the ground with legs crossed, or to eat the lumpy *nshima* I made. I hired her children to sweep my yard and beat away the snakes; she taught me how to clean my pots with mud and find fruit in the forest.

I first met her when I biked by her home on the village path. I'd moved into my own hut at the end of the path. For a week I had settled in, arranged my few books over and over, gotten used to taking splash baths with minuscule amounts of water and cooking over an open fire. Finally, I ventured out into the surrounding area, cruising the new countryside on my shiny, Peace Corps–approved, black mountain bike. It is possible Beatrice had waited all week for me to come by. When I biked past her home, she came running out, yelling, "*batoka muzungu!*" which roughly translates to "blond white person." I stopped my bike, and suddenly, this large woman with a huge smile whom I did not know hugged me, handed me a gift of spinach, and invited me to dinner the next day. All this happened even though we did not speak a common language. Instant friendship.

I wrote home about her, telling my own mother about this Zambian mother, who, just two years older than me, already had a husband working for the copper mines and seven children.

The day of the second Kaonde funeral, we sat together, she and I, in the shade of her hut. When the wailing started, Beatrice sent one of her sons down the path, and he returned quickly, speaking so rapidly in Kaonde that I could not understand. I was still learning the language, and caught about every third word. Beatrice shook her head at her son's words and looked at me, smiling. I was not reassured, because I'd learned that often Kaonde people will smile at anything.

She told me, in broken English, that they'd been drinking, the funeral goers, that the coffin was coming, and that it was witchcraft. She yelled, and the children vanished. And then the wailing was much louder. They were upon us.

A crowd of men and women, billowing dust, and a dizzying array of bright fabrics and sweat and dogs appeared on the village path and poured into Beatrice's yard. The coffin floated on rotating fingers while the women wailed. It was deafening.

The crowd brought the coffin around and surged up to me—just one second—and I was nose to nose with a rough-hewn wooden box. It was pulled away, then it moved to Beatrice. The bearers didn't have as good a control over the coffin when they confronted Beatrice; the inertia of the crowd smacked the coffin into the side of her head, bloodying her. She fell to the ground and the crowd engulfed her and I screamed, but then, Beatrice was magically standing to one side as they all swirled and drained away.

One woman I saw raked fingernails down her own face, drawing bright blood. She was beyond this place, this location—all of them were, moving, revolving, releasing such emotion for the death of this person stuffed into a small, brown box. And then they were gone, weaving down the path, away from us, leaving behind a cloud of fumes and thick dust.

Beatrice wasn't hurt badly. Her ear bore a large cut that I patched with Peace Corps Band-Aids. Her children leapfrogged in the yard. I washed the dust away, and her blood.

The wailers, she told me later, much later, when I understood Kaonde better and she understood English, were embodying grief and sadness and expressing it for all to see. Women will wail for people they know and people they do not know. It depends on how pent up their emotions are. Often, the coffin of a dead person who is suspected of dying by witchcraft will lead mourners to the killer. How mourners know someone was killed by witchcraft depends on how the coffin acts at the funeral. The logic is circular.

I never wailed. Several villagers died while I lived in Zambia, and I attended their funerals, which were somber affairs without the parading of the coffin. Women wailed, but I did not give myself over to wailing.

Every morning in Skagway, when I worked as a guide on the glaciers, I biked to the helicopter base and loaded our gear into one of the helicopters, and then the five of us (Mario,

Kyle, Jason, Elizabeth, and I) piled in with a pilot, and we flew out to a glacier to set up camp for the day.

Some days the ride to work was smooth, soaring over jagged mountain peaks, spying on small glaciers and sun-speckled snow and mountain goats and alpine lakes and blooming mosses and floating eagles. I peered down onto topography defined by pure white, a white that was reflected and blinded the eyes.

Often, the flight would be full of cooling pockets of bumpy air and winds and dense cloud negotiations and dead-end valleys. The pilots wouldn't fly where they couldn't see, so if we tried to maneuver into Sawmill Pass and found it clogged with clouds or fog, we'd have to fly south and try to get inland somewhere else.

The helicopter would pitch up and down, dropping into air pockets, getting bashed with wind and rain. I had faith in the pilots, but faith in fellow humans only gets you so far; more often than not, on wild, rocky rides, I'd turn my faith to the lands outside the helicopter, the blazing stretches of white ice and snow. If a helicopter goes down, there is nothing anyone can do. They are not indestructible, and the land around Skagway was rarely flat and conducive to smooth landings.

Once, a pilot bringing people to the Meade Glacier caught an eagle in his rotors. He emergency-landed, afraid of what would happen to his engine as eagle bits were sucked into his intake. The pilot landed safely, and, hearts pounding, the passengers stepped out onto the ice whole and intact. I remember seeing the helicopter covered in blood, the remains of the eagle splattered across the machine with small white feathers stuck on the bright red panels, glued by the blood. The base was called, a new helicopter flew out to get the passengers, and later, mechanics flew out to the stranded machine and went over it with a proverbial fine-toothed comb. Declared fit, the helicopter was flown safely back to Skagway. Frankly, this is what luck looks like.

Years later, I painted that image. The remains of an eagle spattered across a helicopter. It was imprinted on my eyes, and I needed to release it. I painted a large, blue glacier, sweeping from corner to corner of the canvass. I placed naphthol-red over the blue and, when that dried, single strokes of white.

Grief is a white-hot flame burning in the stomachs of those who remain. Meghan O'Rourke writes that "loss isn't science; it's a human reckoning."

I think about Zambia when I walk the pathway from the barns to my parents' house now. I stop where the moon rocks are piled and always try to clear away the leaf and plant detritus that accumulates on them like lint. There is no one left for me to show the shine of these stones. My mother is not there to scoot along on her bottom, weed the flower bed down the walkway, excitedly call out over early spring crocuses and daffodils.

Three days ago, when I passed the rocks, I stopped, and then, because it felt right, I knelt and plunked down onto the path. I picked through the willow leaves and the crab grass coming up. I gathered the white moonstones individually and wiped each down. They piled up on the pathway, and I counted them. Nine total. There used to be more. I don't know where they all went. I put the remaining ones back, where they belonged, and they sat there, shining white, reflecting.

M Y DAD WAS OFTEN CALLED A GLACIAL SHEPHERD.
My brother and sister and I would tell him this regularly,
whether in person or over the phone. I talked to my dad
every day after my mom died. I was stuck in Missoula,
plugging away at my graduate degree, and he was at our
farm in Washington, building a new life. I couldn't drive
home every weekend, so I phoned. I called him in the evenings, the mornings, I'd bug him in the afternoons and ask
for accounts of his day and what he was working on and
how the animals were, and in the winter, I asked after his
glacier.

We have a pond on the farm, a spring-fed swimming
hole rimmed with willows that stick their rooty toes into
refreshment year-round under the shadow of the windmill. I
learned to swim in that pond as a child. So did my cat.

Originally, my dad dug out the pond around a small
spring to serve as an overflow area for the windmill. Twenty
years or so ago, Dad dug deep into the clay, then built up
the embankment around the pond. The windmill pumped
excess water into the depression, and soon, the pond was
born. Summers filled with my parents' three kids' excavating,
drilling, shaping, and ditching the old levee had caused radical degradation. Reinforcement was desperately needed. Dad

kept having to build up the pond walls and reinforce them with old railroad ties. As kids, we resumed our never-ending remodeling.

Unfortunately, the pond is located at the leeward end of a minor slope. The lay of the land is such that you wouldn't necessarily see, at first, the angle, but come winter or high rains, the water fairly sluices down the surrounding pastures towards the pond. It fills quickly as frogs jump in delight and herons cruise for lunch. Sometimes, I stand in the shelter of the house's deck during a downpour and watch the pond teeter towards complete overflow.

The design flaw, very visible in times of high rains, is that the embankment my dad kept repairing stands over the farm's driveway by a foot or two. Not too high, but enough that, in winter, water seeps through the embankment and floods the loading bays down by the barns. Cold, Pacific Northwest nights freeze the constant runoff. During the day, more water seeps. And freezes. And seeps. You can go down and inspect daily each new, frozen stratum. Technically, the farm glacier was not a real glacier, as it did not persist year after year and was made from ice instead of snow. But such nuance was lost on my childhood, science-oriented brain. I called it a glacier, and soon, we all called it a glacier.

As a child, I learned to ice-skate on the yearly glacier. So did my cat. I had an old pair of sneakers and required only a good running start. I'd hurl myself onto the ice and slide the fifteen feet across. I knew then, as a child, that I was a small step away from an Olympic figure skating career.

My dad would get his sump pump out sometimes and drain the pond down two or three feet. This curbed the glacial growth, but only for a couple of days. It's so damp in the winters that getting ahead of the water is difficult. It seeps. Personally, I liked our glacier, and I think Dad did, too. I'd catch him checking on it, eyeballing the new growth. He had to traverse the ice to get to the barns to feed the cows and do myriad other farm-barn activities. Dad's permanent shadow, our farm dog, Roon Toon Toon, would habitually wipe out on the glacier. Roon Toon Toon was not made for ice-skating; his inner need to herd things was thwarted all winter, as he was forced to circumnavigate the ice.

My brother and sister and I used to tease Dad about his glacier. He'd huff and puff, which just incited more teasing.

Once, with coffee in hand, Dad and I stood down by the barns and assessed the glacier. He was wearing his bright orange rain jacket, his oatmeal-colored hat that Mom had knitted for him tucked over his ears. Dad's white hair stuck out along the fringe. His white, Yosemite Sam mustache was growing icicles under his nostrils. Roon Toon Toon was going in for a lean around Dad's boots.

I monitored the situation. I smiled. "I think, Dad, you'll need to start roping up."

Dad looked at me, peering over his steamed-up glasses. His eyes were gray-blue. "I'll go around."

Once, my dad and I were driving back down the mountain to our farm in Washington. We'd been in Eatonville most of the afternoon, celebrating the sixth birthday of the Mountain Co-Op, where Dad volunteered. He'd started volunteering after Mom had died, trying to fill the hours and days her absence had created. Driving back home, we talked of the people I'd met, of how the town of Eatonville slowly had changed, and of Dad's hopes to supply the co-op with produce from our farm next summer.

Dad drove. He negotiated the twisty mountain roads tilted slightly forward in his seat, his bushy white eyebrows hanging over the steering wheel like antennae. I engaged in the age-old sport of backseat driving, pointing out missed opportunities for blinker use, commenting on recommended highway speeds, suggesting how he could avoid nefarious raindrops.

Coming around a corner, both Dad and I spotted simultaneously the lone figure walking on the right side of the asphalt. Green hood up, cigarette smoke billowing, the man didn't move as our car whipped by him.

Dad and I nurtured a long silence in the Subaru. Then Dad said, "What a waste. That's just too bad."

The young man we drove past was well known in Eatonville. I knew him personally. He got sucked in to drugs early, lured by the siren song of meth.

Dad and I continued home in silence, both of us distracted. Pulling into our driveway, Dad commented again, obviously still thinking of that man: "What a waste. We have so little time."

We revolved around prescriptions of time in my family, especially after my mom's death. Doctors said there was not much time left for Dad, and my siblings and I were only too aware of how quickly time had fled before Mom died.

We, even though we did not talk about it, collectively shifted gears. The moment Mom died, we shifted to keeping Dad alive. And when Dad died, we mourned both parents as one.

We used the language "making the best of the time we have." I heard my brother say this over and over, my sister echoing him on long-distance phone lines.

I remember acutely my mother's wanting more time. More possibility, perhaps a cure or transplant. And so she made the best of the time before her, waiting.

When she lost her leg, she didn't let that curtail her movements. She ranged far and wide on the farm, just not as quickly. As she became sicker, her lungs laboring harder and harder for smaller amounts of oxygen, she didn't give up, didn't turn in her cards to the man beckoning from the doorstep. My mother plumbed her inner depths, summoned up strength and courage, and asked her family to become her legs. If she couldn't go out into the flower gardens anymore, she, we, all of us, brought them to her. In her last years, my dad and my sister and my brother and I packed the decks around our small cedar house: they were flowing and riotous and intricate and woven with flowers and plants and tomatoes and volunteers and hummingbirds and joy.

My sister, Sarah, wears my mother's wedding ring. It's not the ring my mother was married in—that broke years ago—but it is the one she wore on her ring finger for as long as I knew her.

The ring is gold, with one large diamond and two supporting diamonds. They sit deep in the ring. It is a quiet ring.

I was used to seeing that ring on my mother's finger. She wore it, and I never focused on the ring, but on her hands, her fingers. For much of her life the ring was just there, resting lightly, catching light.

My sister has long, white fingers, and it feels like the ring is more prominent somehow. Sarah gestures when she speaks—languidly, soft gestures matching a soft voice—and I find myself staring at her ring. It is as if we are making eye contact, her ring and I. The central diamond throws light and sparkles, winking.

Sarah has my mother's complexion, long, dark hair; brown eyes; a slight build. People always commented that it was easy to see them as a mother-daughter pair. There are times when it feels like Sarah embodies my mother, has inherited not only my mom's physical traits, but also the way my mom inhabited space. When I am in the kitchen with my sister, it is easy to close my eyes and feel like I am with my mother.

My brother, Grant, carries my dad's complexion, build, smile. Grant is obviously my dad's son, from the way both move, think, interact with the world. I've seen pictures of my dad when he was young, and they are just like my brother's early pictures. It's the well-over-six-foot build, the blond hair and blue eyes, the booming laugh echoing across the farmyard.

After my parents died, I met my brother and sister at a diner not too far from the farm. We'd been there innumerable times before. My brother ordered his usual egg scramble for dinner; my sister, a chicken sandwich. We sat there, three kids, adults, and talked about the future. What did we want it to be?

I am the youngest, a mixed bridge of my parents. I have my dad's blond coloring and my mother's build, her feet and his big toe. I am fifty-fifty, and sitting in a diner across the table from my two older siblings, it was not a big stretch to imagine sitting with my parents.

In the diner, Grant and Sarah sat across from me and smiled. We ate. In a way, in the years leading up to this moment, we had never actually imagined this happening. Before, it seemed that all time, all imagination, all forward motion flowed around and up to a parental death. After that, things stopped. It had been nearly impossible to imagine what would happen afterwards.

Yet we sat there and talked through the future we imagined. My sister had earned entry into a prestigious graduate program at the University of St. Andrews in Scotland. My brother was expanding his company, investigating buying another competitor. We all three had, in a way, lived through the very worst of our fears, the worst of our imaginings, and survived. My brother's laugh came and went while my sister waved her hands, her ring attracting my magpie eyes.

We each bear tokens of our parents: some more visible to the world than others. My sister's token, her ring, is very visible to me. The diamonds on it intrigue me, draw me in. I carry a complicated fascination with the stones and wonder how my sister is able to handle wearing such complexity upon her finger. Half the time I see the diamond, a crystalline carbon, a beautiful representation of my mother's married life on my sister's finger, and the rest of the time I see reflected back the tangible manifestation of carbon dioxide, one of the very substances that is changing the face of this planet.

My sister said, sitting there, that she was excited to see what the next year would bring. "This year has been so hard," she said. "It will be better next year."

Planetary climatic changes are deeply intertwined with the emission of carbon dioxide, the primary greenhouse gas. Human society releases too much carbon dioxide into the atmosphere, which causes long- and short-wave radiation from the sun to be reflected back and forth between the planet and the atmosphere. Earth's planetary processes are changing because of this. Everything is shifting so rapidly that we, human society, can't keep up. Other greenhouse gases, such as methane, nitrous oxide, ozone, and CFCs, contribute to the climate change phenomenon, but as a society, we don't fixate on them as much. What we do talk about is carbon, carbon, carbon.

Carbon is found everywhere on Earth, in the sun and stars and comets, on Gliese 581g, and out in the far reaches of space. Carbon is the fourth-most-abundant element; it is found in all known life. It is the chemical basis of our existence. Carbon is in our atmosphere; in oceans, lakes, and seas; it is in the land and the wind and the food we eat. It dangles from our ears and adorns our engagement rings. Eighteen percent of the human body is carbon. The rest, approximately 80 percent, is H_2O—water. Essentially, we're carbon beings. We breathe in oxygen, breathe out carbon dioxide. We are carbon. Without carbon, what remains?

Agonizingly, what is essential to life is also what is profoundly harming this world. Walt Whitman wrote that humans "contain multitudes." Taken literally, carbon is a large portion of those multitudes.

Here on Earth, there is a finite amount of carbon. Outside of a well-funded research lab, the center of the sun, or perhaps even an atomic bomb, it is nearly impossible to summon the necessary energy to convert one element to another. Carbon is an element, a building block of life. However, carbon has different properties based on how it manifests itself. Think about the difference between a diamond and a chunk of coal. They are not interchangeable, yet they are still carbon in its purest form. The only difference is in how the carbon manifests itself. For example, you would be much more likely to reach for a chunk of coal to burn than a diamond gem.

We are taught in school the wonders of the carbon cycle, how carbon is recycled through all planetary organisms and the biosphere. It is planetary currency for life, whether in the form of crude oil, diamonds, amino acids, or human exhalations.

The Deep Carbon Observatory, a decade-long study, aims to figure out not just what carbon is doing on the surface of Earth, but what carbon does within the planet. Scientists are unsure about carbon's role in the physics of the planet and how much influence the carbon within Earth has upon the carbon on the surface.

A common representation of carbon from deep within the planet is a diamond, which I envision to be the equivalent of a planetary eureka. A diamond's inception occurs at depths ranging from eighty to one hundred twenty miles below the surface, in the dark womb of the planet. Carbon is compressed, heated, manipulated into cold hard crystals that surge to the surface and land upon my sister's long, white fingers.

We don't think much, as a people, about carbon, unless it is in forms we can use, in our gas tanks and pencils and plastics and jewelry. It would be a novel world indeed, if suddenly the carbon emissions from our cars and industrial sites would pour forth diamonds rather than gases.

I think in many ways that the decks of my parents' house gave my mother more time to live. She could have left us all much sooner, retreating into the mental confines of pain, sadness, and isolation, if the decks hadn't been there: convenient, accessible conduits to the outdoors. Her oxygen cord was just long enough to give her the freedom to reach the decks, where she would watch the farm and remember her life. My mother breathed and existed in the natural world, with dirt under her fingernails and sunshine in her hair. At her best, she was a glorious sight with her feet planted in the soil.

Dad placed benches on both decks. One deck faced towards Mt. Rainier, the big house garden and the trees and the road and the greenhouse. We called that area "the front." The other faced the back, the barns and the driveway and my dad's quasi-glacier and his shop where he worked metal and wood and other materials. The bench Dad had parked on the front deck suited Mom best, because it had a nice wide berth, room aplenty for her books and her gardening supplies and her family and her. She kept a close eye on the gardens, the apple trees, the flowerpots, the greenhouse, the pump-house, and the pastures, easily viewed from her perch atop that bench. Two folding chairs rested on the far side of the bench. The one closest to the door became my permanent spot.

In the summer before her death, we sat there often. We talked of small things and lapsed into quiets both peaceful and pleasant. I am concerned now with how much wasn't said. I cannot imagine a harder burden to bear: she was ever moving and lively and sparkling and in love with the land. She knew the plants, the nuances of the entire farm, the dips and vales and trees and her children's favorite hiding places. She knew which way the water flowed. To be divorced from that, to slowly have to sit out, to be forced into an observer's position: this burden would have broken lesser women.

Once, I weeded the flower garden that bordered Dad's shop. It took me the entire day to weed, prune, and fight off the invasive Virginia creeper. Exhilarated when I finished, I went into the house, covered in dirt, to triumphantly share my battle, and to report the survival of the daffodil bulbs and the discovery of what surely must have been a rabbit castle. My mother was having a bad air day and her breathing was labored. She sat at the kitchen table, looking out the window. I, in the kitchen, washing my hands and nattering on about the flower garden. She, silent. She finally said something; I don't recall what. But it opened the door to a fight, and suddenly I was yelling at her. It happened so quickly. She was throwing angry, bitter words at me, and I was yelling back. I remember asking her specifically what I should do to make her happy: not work the gardens and let the weeds overrun them, or keep on caring for them?

Even then, I knew the fight was not about the flower garden. She wanted to be outside with me, sharing the battle against the Virginia creeper. She wanted to resuscitate her daffodil bulbs. Such anger is so understandable. In the face of unimaginable challenges, anger is a natural response. But it cannot be a permanent state. I recoiled from her, and she from me, and an hour, two, three passed, and then we were sitting together on the front porch again, spotting humming-birds, having one of our quiets.

My dad and I spoke one night on the phone while I was in Missoula, taking a pause from the ever-growing pile of work associated with graduate studies. We talked about Mom, about her strength. My dad recalled the time right after she was discharged from the hospital following the am-putation of her leg. My dad was working on Kodiak Island in Alaska, and my mother flew up there to take the time to recover.

"Kodiak is small," Dad told me. "It became the thing every day that no one would borrow my work truck in the afternoon. I would leave it parked, and your mom would take it down to the fishing docks."

"The fishing docks?" I asked.

"Yes. She taught herself to walk again. She put the time in. For hours, up and down, she walked the piers."

The community in Kodiak was small, protective, and soon people started talking about the woman staggering up and down the waterfront, hand on the rail, lurching on her prosthesis.

"I would get two, three calls in the afternoon. She fell down. But the guys helped her up. It happened every day."

"Would you go help her?" I asked. I already knew the answer. My dad had told me this story many times. We both drew strength from it.

"No. Not in the afternoons. She wanted to teach herself how to do it, how to walk. On the oily, fishy, uneven, wooden planks, she'd go back and forth."

Physical therapy for my mother involved rough-hewn, oil- and fish-slicked boards, with falls that could topple her into cold ocean water. I imagine how that must have felt for her, to be running on nothing but sheer strength reserves and force of will, winds carrying the smells of the Pacific blasting icily through her hair. She could have done outpatient physical therapy in Washington State. But instead she boarded a plane, flew to a remote corner of Alaska, and did it herself. The glaciers sat, perched in their high mountain saddles, and watched, knowing the inherent courage and imagination found in time invested, force of will enacted.

Today, there is a fixation on time in climate change discussions. It's in the literature, the magazines, the news, the IPCC reports. How much time is there? Years? Decades? Centuries? Climate models are humming, spitting out dire predictions. If human societies do not collectively respond to climate change very soon, then this variegated, goldilocks climate we live in will be altered drastically across the planet—some fear that we may no longer be able to survive. Soon, even if individuals and societies and organizations and governments put aside their myopia and join together, even if, globally, we can agree and subsequently reduce greenhouse gas emissions, even if we improve the technology needed to create feasible, clean energy, even if people choose to start seeing and making meaning of and acting upon the underlying social inequities that climate change highlights, even, even, even—the inertia of climate change might be too much. The cumulative effects of climate change are simultaneously gradual and quick, manifesting on timescales ranging from instantaneous to long term. Climate change could build upon itself until it becomes a planetary wrecking ball.

The answer to climate change is simple—reduce emissions, utilize clean energy—yet terrifying in its complexity. There are vast numbers of social, cultural, and natural systems bound up with those simple answers. But there is not enough time to demand perfect solutions for reducing emissions and utilizing clean energy. In 2008, the World Glacier Monitoring Service announced that glaciers were melting more quickly than at any other time in the last five thousand years. Once a glacier is gone, it's gone, regardless of how much white paint is carefully placed on a mountainside.

I feel, at times, like my dad's fish, Albert, swimming in his water trough. Only, instead of where he is now, lapping the pool and dodging the occasional cow nose, it feels like the trough is getting smaller; the ice, thicker. Albert has the same problem as humanity, climate change, but with different details. His is a problem of cold, of too much ice; ours is of heat, of increasing temperatures. Options and choices are growing fewer. I can't leave the tank—Gliese 581g sure looks nice along that one narrow strip of perfect temperatures, aside from it being so far away—but I also can't fix the tank. Not alone.

When my mother needed to learn how to walk again, instead of doing a step-by-step program, she summoned her courage, one plastic footstep on oily planks at a time. I take inspiration from her and admire her strength. Yet I want to build on it, add to it. I cannot help but think that we need to engage willfully and imaginatively with climate change now, together, as a people who collectively want to imagine a better home. We need to slip on the oil and trust our fellow friends and family and strangers to help us up; we need to feel the wind in our hair and reach deep for strength. I do not think we have the time, the luxury, to wait for perfection. Rather, I think we might be better served by diving in, getting messy, and trying over and over as we fight our way forward. We need the graphite and the diamonds, the dark and the light.

There are times that I do not believe that my mother died, in my mother's death, just as there are times that I do not believe that climate change is real. I simply could not imagine them. Neither of these things, however, requires my belief. They are both realities.

Ken Kesey writes, in *Sometimes a Great Notion*, that "time overlaps itself. A breath breathed from a passing breeze is not the whole wind, neither is it just the last of what has passed and the first of what will come, but is more—let me see—more like a single point plucked on a single strand of a vast spider web of winds, setting the whole scene atingle."

Kesey understood the complexities of time, the dizzying awareness of living in a single moment, with a mind pointed towards a dreaded future full of the direst of imaginings. I feel often that I am living in many moments at once, summoning memories and futures, and running a wet washcloth over my dad's brow. I am aware that time is overlapping in the months ahead. I know that my father's death will go the same as my mother's did, but also very differently.

In her last years, my mother's daily life became even more difficult. Her lungs labored harder and harder for smaller amounts of oxygen—but she didn't stop fighting. She plumbed her inner depths. She started moving slower. Then her lungs became sluggish, unable, inch by inch, to process oxygen. She couldn't get enough air. It was a gradual process: she started using oxygen in her sleep, and then she used it all day. First, a small dose; soon, the highest dose. We kept the oxygen tanks upstairs and hung the clear plastic tube from the railing. It was her leash: the tube followed her everywhere.

I am afraid, at times, that my lasting memories of her will be of that plastic tubing. I always knew where she was: I just followed the tubing. It stretched thirty feet. This allowed her to move from her room to the kitchen to the living room and outside to either deck. If she wanted to range farther, she had to saddle up her portable oxygen tank, switch out nasal cannulas—but by the time that show was over, she'd be out of air, exhausted, and wouldn't go outside after all.

My dad's situation is different; he is different. On the refrigerator at the farm are papers we must hand the medics if they come responding. The week after doctors told Dad that there wasn't anything more they could do, he signed those papers. No extraordinary measures. No CPR. No tubes. No life support. He made sure my brother and sister and I all knew this. The time he has been given is the time he has. We're on board. We support our father in the decisions he makes with us.

My dad has months left. This is not a slow, dwindling process of gradual suffocation. It is a voracious cancer, and without mercy. We have a team of doctors: they will help with his pain, our pain.

I am reminded of my friend Bruce Schindler, an ivory carver in Skagway. Bruce works, mainly by hand, with wooly mammoth and walrus ivory. I've seen him create many different designs, but where he fights fate is in his wooly mammoth tusk restoration. He makes the tattoos of time, the glacial cracks in the landscape of the tusks, disappear. Mammoths walked the Earth for 2.5 million years before going extinct, and their wanderings took them all over the globe, especially to the high north. Today, the thawing northern tundra yields tusks that are relatively intact, preserved for more than 35,000 years.

In Bruce's workroom, he hangs the ten- to-fifteen-foot-long tusks with climbing rope from specialized roof mounts. Before he begins working on the tusks, they're gray-brown, brittle, and cracked—they emit an odor of oldness. Frankly, it is hard to stare at these drab things and fully imagine the future Bruce breathes into them.

Bruce will work with them, a time-intensive process, sanding and smoothing and stabilizing and flicking and staring and inspecting. He'll navigate and isolate each individual crack in a tusk, filling it with clear or black epoxy and coating the entire tusk with protective varnish. It is incredible to watch an aged thing be reborn strong and sturdy. Weeks, months later, Bruce will call me up to visit his studio, and I will see the end product of his slow, imaginative work. I see the tusks as they were and imagine the magnificent creatures that bore such immense defenses.

I find that I prefer it when Bruce uses black epoxy to fill the deep cracks in the tusks. Some cracks run the entire length of a tusk, ten feet or more. The black contrasts sharply with the tusk itself, which, to me, acts as a reminder of age, of time, of imperfection, of flaws in the original whole. It is easy to look at some of Bruce's restored tusks and imagine that the mammoth that grew them was alive moments ago. But the tusks with the black epoxy—these come from a creature long gone from our world. Bruce has filled in the cracks of time, but he has not erased them. Rather, in these tusks, time is something to wonder at, explore, revere, to run your hands up and down their smoothness.

The remaining months with my dad are going to wink by, mere seconds on the continuum. But it will be like diamond-time, epoxying the months that I can keep, hold precious and close. I don't seek to skim over the coming months, make them disappear in the magic of a life restored, retold. Rather, I need these moments to act as that black epoxy, to fill in the cracks of time and experiences. Robert Brault wrote, "For centuries, man believed that the sun revolved around the earth. Centuries later, he still thinks that time moves clockwise."

We will tend the cows, work the fields. I am growing many different vegetables and flowers; my father is growing his glacier again. When winter finally finishes and spring returns, we will turn our faces to the sun and revel in its warmth. I will stand witness to my dad's last days here on this planet, just as he witnessed my mother's. In a way, the bracelets I wear, the diamond ring upon my sister's finger, my brother's laugh—these are in themselves witnesses to lives lived, even as carbon—the essence of life—is, itself, a witness.

My dad nearly missed the turn onto our road when we came back down from Eatonville. The cancer sometimes clouds his brain. He still has full mental capacity, but there are also moments when he can't remember or draws a blank. We wait for the clouds to recede, for the sunshine to burst through the gray, and then I point out our turn, and he turns, and there's the farm, and we pull into the driveway and see the windmill and drive across the glacier and park and we're okay and Roon Toon Toon is wagging his tail, happy to see us.

I get scared. Individual action in the face of cancer is terrifying. My dad runs a farm and raises cows and grows produce and volunteers at the Mountain Co-Op and helps his neighbors and tends my cat and keeps accidentally checking the "married" box on all the various paperwork he gets and calls his three children and dreams about seeing my mother again and chooses to get out of bed every day.

We can't do this alone, so friends from my life, my brother's, my sister's, my parents', appear, and they pour endless cups of coffee. They hug and they help. Friends help us split cords and cords of wood, cook dinners and desserts for my dad. They take the time to come help us hay in the summer, chase our cows, remember our mother, drag Dad out of the house.

Dad deals with the uncertainty, the frustrations, and the dehumanizing effects of cancer with grace and humor. Dad hates dwelling on what hasn't worked, on what went wrong, on what might have caused his cancer. He's proud that all his children are in graduate school, that we're each doing the best we can by him and each other. The other day, when he and I sat down to calculate how much seed to buy, he stared and stared at the calculator in his hand. He couldn't remember how to use it. The man who used to do trigonometry in his head.

We know what the problem is. We don't have a solution, or answers. There are some temporary measures, some promising ideas, some huge forward leaps and a few crippling losses. But the take-home message here is that Dad hasn't given up, nor has my family. Nor have our friends. Dad is our home, and our home is threatened. He is our past and our future. We're busy with the business of living, the business of moving forward.

I don't think we can remove climate change completely from human society's fate. But we can lessen it, mitigate it, ameliorate it through collective, imaginative action. We can't keep asking for time: there isn't any.

My dad and I walk up the path from the barns to the house. Roon Toon Toon orbits around us, tail pluming. He's happy that we're home from the co-op. Dad gives me a hug.

"Thank you," he says.

"For what?" I ask.

"For doing what you're doing."

We pause by his shop and gaze over to the pond, the windmill, the almost-glacier. The glacier is small now, clumped around the embankment. But it is still there, clinging on. We both look up, beyond it, with honed, Didion-style attention to the slopes of Mt. Rainier, where the real glaciers tumble and move magnificently. It is beautiful to see, a stunning reminder that the landscape is a great foil for the human spirit. We turn, walk the path, and go into the house.

Epilogue

I DID NOT WITNESS IT, BUT I CAN SEE IT IN MY MIND clearly: my father, haggard after lung surgery, upright and tall, slow dancing with his nurse in the ICU. The pair revolves steadily, the tubes and cords and monitors attached to my father allowing them a one-foot radius.

I walked in later, after the dance, when my father was perched on the side of his bed, laboring for breath. The nurse was in tears. Her words to me were simple: "You have an amazing father."

Much later, after she'd gone, I asked my father what had happened. He told me, then my brother, and a day afterwards, he told my sister. The story turned, spilled, but its core remained the same.

The nurse had come in to check on my father. They got to talking, and she mentioned how nice it was to see him surrounded by his children. Apparently the conversation moved, tripped, stumbled, and she volunteered that she'd never had a real father—that she'd been a nurse for eighteen years and was the mother of two small children, but her own family still vocalized their opinion that she had amounted to nothing.

"Ah! But look at what you've done, what you've accomplished!" I know my father said this because when the story is repeated, told again and again in my family, this dialogue remains the same, unaltered.

They talked longer, and then my father, for unknown reasons, reached his hand out to his nurse. "Would you like to dance?" he asked her.

She paused. I imagine the look on her face, a mixture of puzzlement and concern. Not twelve hours earlier doctors had drilled into this man's chest, drained away over two liters of bloody fluid, and then cemented his lungs to his chest wall with talcum powder.

"No one has ever asked me to dance before."

"Come on." My father shuffled out of bed.

They danced.

I know what this moment must have looked like. I can imagine it clearly in my mind.

This single moment tells me that even when we are at our most beleaguered, we are still capable of making this world a better place. That even when the way before us is dark; when we feel the cards are stacked against us; when we're left with nothing but temporary measures to ease the pain; when, ultimately, we feel powerless—we are still incredibly, miraculously capable of action. We are audacious.

Notes and Sources

THIS BOOK IS INFORMED BY AN ENORMOUS ARRAY of people, places, organizations, and experiences throughout my life. Ever-winding, ongoing conversations with scientists, artists, writers, and other glacier-lovers ambled into this story and my dreams in surprising and delightful ways.

Climatic changes occur at multiple spatial and temporal scales across the planet with varying and conflicting implications on the human and more-than-human world. More climate change information, understanding, and nuance is revealed and shared each day, and, to some degree, by the time you hold this book, some information will already be out of date. I encourage you, therefore, to stay up to date, to keep reading, and, most importantly, to share your own climate change stories widely.

Collected here are online sources and additional readings that influenced this book and are exceedingly useful for further educating yourself about climate change.

Online sources: the Intergovernmental Panel on Climate Change (IPCC), ipcc.ch; *The New York Times*, nytimes.com; *Scientific American*, scientificamerican.com; the National Aeronautics and Space Administration (NASA), nasa.gov; The World Glacier Monitoring Service, wgms.ch; National

Snow and Ice Data Center, nsidc.org; National Geographic, nationalgeographic.com; National Climate Assessment, nca2014.globalchange.gov; The Canary Project, canary-project.org; and the Extreme Ice Survey, extremeicesurvey.org.

Academic journals: *Nature Climate Change, WIRES Climate Change, Polar Geography, Global Environmental Change, Polar Research Journal,* and the *International Journal of Glaciology.*

The following books were invaluable as I wrote *While Glaciers Slept*: Kim Heacox, *The Only Kayak*, Lyons Press, 2006; Mark Carey, *In the Shadow of Melting Glaciers*, Oxford University Press, 2010; Susan Sontag, *Regarding the Pain of Others*, Farrar, Straus and Giroux, 2003; E. C. Pielou, *After the Ice Age*, University of Chicago Press, 1992; Rebecca Solnit, *Hope in the Dark*, Nation Books, 2005; Henry Pollack, *A World Without Ice*, Avery, 2009; Julie Cruikshank, *Do Glaciers Listen?*, University of Washington Press, 2010; Barbara Kingsolver, *Prodigal Summer*, Harper Perennial, 2001; Keith H. Basso, *Wisdom Sits in Places*, University of New Mexico Press, 1996; Mike Hulme, *Why We Disagree About Climate Change*, Cambridge University Press, 2009; Keri Hulme, *The Bone People*, Penguin Books, 1986; Andri Snær Magnason, *Dreamland*, Edda UK Ltd., 2012; Barry Lopez, *Arctic Dreams*, Vintage, 2001; Lucian Boia, *The Weather in the Imagination*, Reaktion Books, 2005; and Joan Didion, *A Book of Common Prayer*, Vintage, 1995; *The Year of Magical Thinking*, Vintage, 2007.

Acknowledgments

I AM LEARNING TO LIVE AND LEARNING TO BE OF SERVICE. I am passionately, audaciously inspired to be alive today, to be a part of this growing, thriving, breathing, stimulating, perplexing earth. I am grateful to be supported by strong, proud, vibrant men and women who love and encourage me—during the heartbreaking moments in my life and the stargazingly happy moments.

I am grateful to many people and institutions for their continuing support and encouragement. At the University of Montana, the EVST Program created a safe space for me to rebuild and write. Phil Condon, Dan Spencer, Neva Hassanein, and Judy Blunt guided, inspired, and supported me through countless drafts over countless coffees and bowls of soup. I'm grateful to the musical specialty group More Banjos Less Pants, especially Kjerstin Gurda, Adam Lohrmann, Alex and Peter Carr, Bri Ewert, Zach Johnson, Micah Sewell, Keema Waterfield, and Zachary Carlsen. For walks on windy paths in the Missoula hills, I owe a debt of sanity to Nathaniel Miller and his coyotedog, Lalo.

I have met many different glaciers through the US Fulbright Commission and the National Geographic Society. My thanks to those whose help and kindness made my travels possible. Special mention goes to an especially energetic, passionate, and world-changing group of Narvins.

At the University of Oregon, I am grateful to my doctoral committee, Alexander Murphy, Mark Carey, Andrew Marcus, Shaul Cohen, and Ford Cochran.

This book found an incredible home at Green Writers Press under the guidance of Dede Cummings and the sharp editorial eyes of John Tiholz, Rosanne Alexandre-Leach and Ellen Keelan, along with international student intern from Bennington College, An Nguyen. Thank you for believing in me and in this book.

My gratitude to the unwavering group of lifelong friends without whom I could not have told this story: Kevin Malgesini, Aniela Pendergast, Jordan and Thomas Vroom, Mario Taffera, Erin Houtsma, Megan Davis, and Natalie Hunt. I'm grateful to Jon Marshall for daily support, love, and endless morning coffee.

I am deeply moved by the love and kindness of my family and constellating family friends. Sarah and Grant Jackson. The Naegles. Lyn, Brie, Ali, Jass, Chase, and Meesook Jackson. Becky and Steve Evans. Patricia Payne. Miki and Brian Welfringer. Juli and Dave Warburton. Elizabeth Neuhalfen. Chalon and Rich Bearteet. David and Mac Davey. Cathy Ivers. Al and Barb Svinth. The Grabers and the Rohrs. I am thankful to all of you, and so grateful we together knew and remember the incredible people who were John and Cathy Jackson.